Consolidating Active and Reserve Component Training Infrastructure

John F. Schank
John D. Winkler
Michael G. Mattock
Michael G. Shanley
James C. Crowley
Laurie L. McDonald
Rodger A. Madison

Prepared for the
United States Army

Arroyo Center

RAND

Approved for public release; distribution unlimited

U
408.3
.C665
1999

For more information on the RAND Arroyo Center,
contact the Director of Operations, (310) 393-0411,
extension 6500, or visit the Arroyo Center's Web site at
http://www.rand.org/organization/ard/

PREFACE

This report presents the results of a research project entitled "Evolution of the Total Army School System." The project examines ways to consolidate training infrastructure and augment capabilities across Army components to gain efficiency and achieve economies of scale in conducting individual training of Active Component (AC) and Reserve Component (RC) soldiers. It provides a quantitative approach for determining how the Army might benefit from such changes as offering reclassification training and noncommissioned officer (NCO) education to AC soldiers at RC schools and additional training courses to RC soldiers at AC schools, using the area of maintenance training as an example. If the Army found these benefits worth pursuing, this approach could be adapted and extended to support policy decisions to further integrate its training infrastructure in additional functional areas.

The research reported here was sponsored by the Commanding General of the U.S. Army Combined Arms Center and was conducted in RAND Arroyo Center's Manpower and Training Program. The Arroyo Center is a federally funded research and development center sponsored by the United States Army.

CONTENTS

Preface iii

Figures vii

Tables ix

Summary xi

Acknowledgments xv

Abbreviations xvii

Chapter One
   INTRODUCTION 1
   Background 1
   Objectives 3
   Scope 4
   Organization of This Document 6

Chapter Two
   DEVELOPING AN OPTIMIZATION MODEL FOR
   ANALYZING INTEGRATION OPTIONS 7
   How the Optimization Model Works 7
   Model Database 9
   Model Assumptions 9

Chapter Three
   RESULTS OF ANALYZING THREE
   INTEGRATION OPTIONS 11
   Option 1: Nearest School 11
      Sending Students to Nearest School Results in
      Significant Flows Across Components 12

RTS-Ms Assume Larger Workloads Given Longer AC
    Course Lengths 16
  Sending Students to Nearest Schools Reduces Travel
    Cost and AC Student Time Away from Home 17
  Option 2: Reassign Courses 19
    AC Schools and RTS-Ms Will Have Increased Course
      Offerings, Less So at Specialized RTS-Ms 20
    Student Flows Across Similar Component Boundaries
      in Similar Numbers 21
    Travel Costs Decrease in All Options 23
    AC Student Time Away from Home Also Decreases in
      All Options 23
  Option 3: Consolidate Schools 26
    Results Mirror Those for Previous Options 27
    Number of RTS-Ms Can Be Reduced 29
    Even with Fewer RTS-Ms, the System Is Still Robust ... 30
    RTS-Ms Have the Capacity to Assume New Missions ... 32

Chapter Four
  CONCLUSIONS AND IMPLICATIONS 33
  AC/RC School Integration Provides a Range of Personnel-
    Related Benefits 33
    Instructor Benefits 33
    Support Personnel Benefits 34
    Student Benefits 34
  Results Can Be Generalized Beyond Maintenance
    Training Area 35
  Pilot-Testing AC/RC Integration and Conducting More
    Detailed Analyses of Instructor Requirements Are
    Logical Next Steps 35

Appendix

A. TECHNICAL DESCRIPTION OF THE
   OPTIMIZATION MODEL 37

B. COURSES TAUGHT AT EACH SCHOOL FOR THE
   VARIOUS OPTIONS 43

C. SCHOOLS OFFERING SPECIFIC COURSES FOR THE
   VARIOUS OPTIONS 89

References 131

FIGURES

S.1.	Relative Impact on Travel Costs	xiii
S.2.	Relative Active Component Student Time Away from Home	xiii
3.1.	Flows for Sending Students in Option 1	12
3.2.	Effect on Training Workload at AC Schools and RC Schools in Option 1	17
3.3.	Impact on Travel Cost and AC Time Away from Home in Option 1	18
3.4.	Number of RC Course Offerings at AC Schools and AC Offerings at RTS-Ms for Option 2 Compared to Option 1	21
3.5.	Student Flows of RC Students to AC Schools and AC Students to RTS-Ms for Option 2 Compared to Option 1	22
3.6.	Impact of Travel Costs for Option 2 Compared to Option 1	25
3.7.	AC Student Time Away from Home for Option 2 Compared to Option 1	26
3.8.	Student Flows of RC Students to AC Schools and AC Students to RTS-Ms for Option 3 Compared to Options 1 and 2	27
3.9.	Impact of Travel Cost for Option 3 Compared to Options 1 and 2	29
3.10.	AC Student Time Away from Home for Option 3 Compared to Options 1 and 2	30

TABLES

2.1.	Number of Maintenance-Related Courses	9
3.1.	Specific AC and RC Student Flows in Option 1	14
3.2.	Number of Schools Offering Selected Courses: Option 1	15
3.3.	Breakdown of Travel Costs for Baseline and Nearest School Options	19
3.4.	Number of Schools Offering Specific Courses: Option 2	22
3.5.	Specific AC and RC Student Flows in Option 2 (Specialized Case)	24
3.6.	Breakdown of Travel Cost Savings for Option 2	25
3.7.	Specific AC and RC Student Flows in Option 3 (Specialized Case)	28
3.8.	Number of Schools Offering Selected Courses: Option 3	31
A.1.	Relationship Between Groups and MOSs	38

SUMMARY

The recently established Total Army School System (TASS) has made some significant changes in how the Army provides individual training, but to this point it has concentrated primarily on changing how Reserve Component (RC) training institutions are organized and managed. This is leading to improvements in the efficiency and performance of RC schools.[1] However, given continued pressure on training resources and continued evidence of underutilized training capacity in the active and reserve components, there remains a need to improve training resource utilization and potentially reduce training infrastructure across the entire Army. Such improvements may be possible through further consolidation and integration of training resources into a common Army training system that is truly seamless and "component-blind."

The objective of this analysis is to understand the feasibility and potential benefits of further integrating Active Component (AC) and RC schools offering reclassification training and noncommissioned officer (NCO) education. To meet this objective, we developed an optimization model that determines the least-cost assignment of students and courses to schools under a variety of options. We looked at three options in the area of maintenance-related training, focusing on RC Regional Training Sites–Maintenance (RTS-Ms) and the AC proponent schools offering maintenance courses (primarily, but not exclusively, Aberdeen Proving Ground). The premise we examine is that an accredited Army school with certified instructors

---

[1] For a discussion of TASS implementation see Winkler et al. (MR-955-A, forthcoming).

can offer maintenance courses to any soldier (AC or RC) in its local area.

The options that we analyzed were as follows:

- **Nearest school.** We changed only the location where a student receives training, holding fixed (at fiscal year 1996 levels) the schools and the courses offered at each school.

- **Reassign courses.** We assigned AC and RC maintenance courses to AC schools and RTS-Ms based on local demand. Here, we considered two cases, one where RTS-Ms taught a wide range of courses (multifunctional) and one where they specialized in a specific functional area (specialized).

- **Consolidate schools.** We examined the number of schools that offer maintenance-related training if Army maintenance schools offer courses based on localized demand. For this option, we concentrated only on the "specialized" case described above.

Our exploratory analyses suggest that permitting AC and RC students to take the maintenance courses they need at the nearest accredited school (AC school or RTS-M) is feasible and offers a range of economic and cultural benefits. Travel, per diem, and (potentially) instructor[2] costs are reduced by allowing AC and RC students to take the course they need at the accredited Army school nearest to their home location independent of the component of the student or the school. For example, Figure S.1 shows that travel costs are lower for each of our options compared to a comparable estimate of travel costs for fiscal year 1996, the period we examined. Likewise, Figure S.2 shows the time AC soldiers spend away from their home location for the training, and, therefore, per diem costs are reduced under each option.

Morale benefits include a reduction in the time AC students spend away from their homes and their units and a reduction in the train-

---

[2]In our analysis, we did not specifically examine the requirement for, or the use of, different types of instructors at the AC schools or the RTS-Ms. Assigning courses and students to schools under the options we considered might allow more optimum use of the AC and RC instructors. This is certainly an area that warrants further research.

Summary xiii

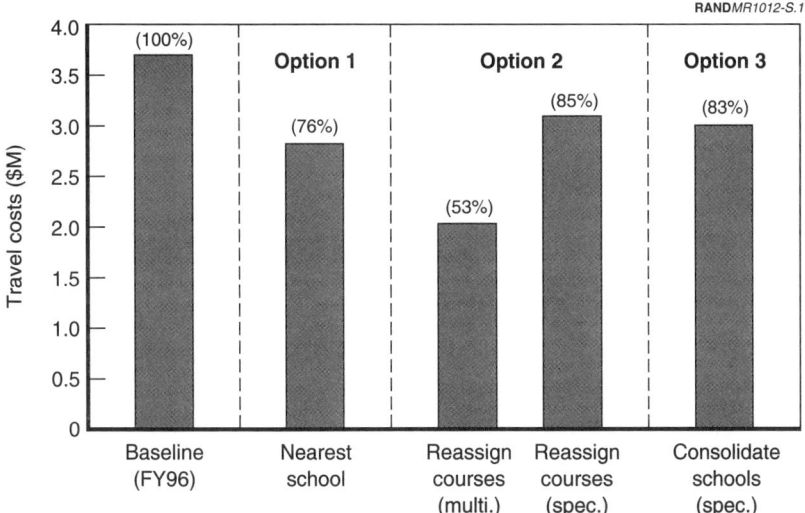

Figure S.1—Relative Impact on Travel Costs

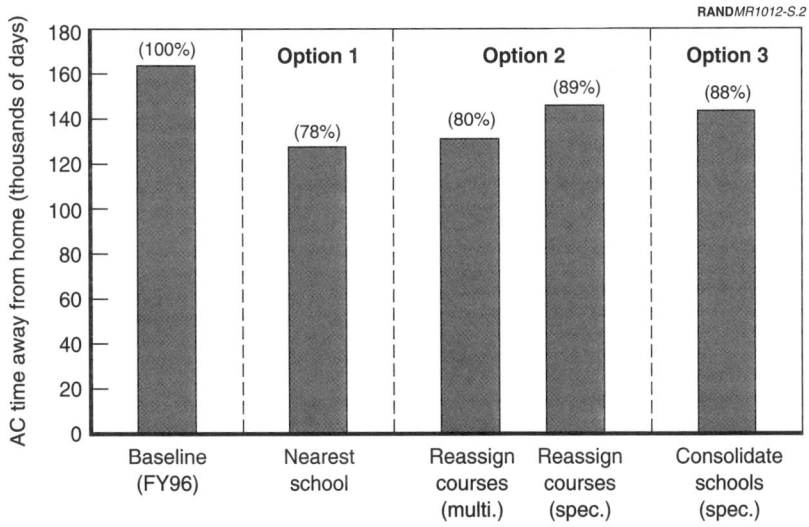

Figure S.2—Relative Active Component Student Time Away from Home

ing workloads of AC instructors. More important, integration of AC and RC training can increase interaction and potentially build trust and confidence across components.

We have examined maintenance-related courses, but there is no reason the types of integration we are proposing could not also work in other areas, particularly where RC schools with fixed facilities, organic equipment, and full-time staff exist or can be made available (i.e., where RC regional training sites have been established). Such Combat Service Support (CSS) functional areas as medical, transportation, or quartermaster are obvious extensions, as are such Combat Support (CS) areas as military intelligence, engineering, and aviation maintenance.

RTS-Ms can also play a valuable role with the advent of distance learning. Local RTS-Ms can provide facilities, and instructor expertise, to help both AC and RC soldiers who are taking courses, or portions of courses, outside a resident school environment.

Based on these findings, we recommend a pilot test to better understand the options and the policy implications of integrating the AC and RC training systems, focusing on maintenance and transportation. In addition, we recommend more thoroughly examining the instructor requirements that would result from a more fully integrated school system.

ACKNOWLEDGMENTS

The authors benefited from support and assistance provided by many people in the U.S. Army. We owe particular thanks to Lieutenant General Leonard D. Holder (ret.), who sponsored this work when he served as the Commanding General, U.S. Army Combined Arms Command. In addition, key support and assistance was provided by staff at the Army National Guard, Office of the Chief of the Army Reserve (OCAR), Office of the Deputy Chief of Staff for Personnel (ODCSPER), U.S. Army Training and Doctrine Command (TRADOC), U.S. Army Combined Arms Support Command (CASCOM), the U.S. Army Ordnance Center and School, and various Regional Training Sites for Maintenance in the United States.

We are also grateful for assistance received from our RAND colleagues. Al Robbert and Herbert J. Shukiar provided helpful technical reviews of an earlier draft of this report. Paul Steinberg provided very valuable assistance as a communications analyst, and Molly Coleman prepared documents and briefings based on this work. While we have benefited greatly from assistance provided by all these sources, any errors of fact or interpretation remain the authors' responsibility.

ABBREVIATIONS

AC	Active Component
ADT	Active Duty for Training
AGR	Active Guard and Reserve
AIT	Advanced Individual Training
ANCOC	Advanced Noncommissioned Officer Course
ARNG	U.S. Army National Guard
ASI	Additional Skill Identifier
AT	Annual Training
ATRRS	Army Training Requirements and Resources System
BNCOC	Basic Noncommissioned Officer Course
CA	Combat Arms
CS	Combat Support
CSS	Combat Service Support
DMOSQ	Duty MOS Qualified
FY	Fiscal Year
IDT	Individual Duty for Training
IET	Initial Entry Training
MOS	Military Occupational Specialty

NCO	Noncommissioned Officer
NCOES	NCO Education System
RC	Reserve Component
RTS-M	Regional Training Site–Maintenance
RTS-T	Regional Training Site–Transportation
SIDPERS	Standard Installation/Division Personnel System
TASS	Total Army School System
TATS	Total Army Training Structure
TDA	Table of Distribution and Allowances
TDY	Temporary Duty
TRADOC	U.S. Army Training and Doctrine Command
USAR	U.S. Army Reserve

Chapter One

INTRODUCTION

BACKGROUND

Historically, the Army ran three separate school systems—one each for the Active Component (AC), the Army National Guard (ARNG), and the United States Army Reserve (USAR). In 1994, the Chief of Staff of the Army, realizing the need to improve quality and increase efficiency in the face of declining training resources, called for an integrated training system to serve soldiers of both the AC and the Reserve Component (RC). This new concept was referred to as the Total Army School System (TASS). Intended to improve quality, efficiency, and performance, TASS involved organizing the nation into separate regions, consolidating existing RC training institutions, and making one organization (and one component) responsible for managing training in a single functional area (combat arms (CA), combat support (CS), and combat service support (CSS)). In addition, within the TASS concept, the RC schools were to be linked to their AC counterpart proponent schools, which would be responsible for quality assurance by accrediting the RC schools and certifying the instructors. The TASS concept was initially tested as a prototype in one region—Region C—the southeastern United States, which encompasses North Carolina, South Carolina, Georgia, and Florida.

TASS has made some significant advances that are improving the efficiency and performance of RC schools.[1] However, to this point,

---

[1] For a discussion of TASS performance, see Winkler et al. (1996), Shanley et al. (1997), Winkler et al. (MR-928-A, forthcoming), and Winkler et al. (MR-955-A, forthcoming).

TASS has concentrated primarily on changing how RC training institutions are organized and managed. Currently, RC soldiers still receive training nearly exclusively at RC training institutions, and AC soldiers receive training at AC training institutions. Given continued pressure on training resources and continued evidence of underutilized training capacity in both the AC and the RC, further improvements are needed in how the Army uses its training infrastructure and resources. Such improvements may be possible from cross-component consolidation and leveraging of training infrastructure and by integrating training resources into a common, "component-blind" Army training system that serves all Army soldiers irrespective of their component.

Such resource integration can have economic, efficiency, and cultural benefits. In terms of the economic benefits, a more fully integrated school system should provide the current level of training at reduced cost. Allowing a soldier to take a course at the Army school closest to his or her home or unit location, regardless of which component currently "owns" that institution, can reduce travel cost and, in the case of AC soldiers, the cost of per diem associated with temporary duty (TDY) training. Economies of scale may result in fewer instructors, or even schools, needed for training.

Such integration can also yield efficiency. With reduced budgets, some of the training missions of AC schools cannot be fully accomplished. For example, under pressure to reduce AC training resources, the Army eliminated the mobile training teams that would go to units to conduct functional and/or new equipment training. Existing RC training assets could help in this regard by assuming some of the auxiliary missions once performed by the AC proponent schools.

Finally, integration can provide cultural benefits. Specifically, a significant benefit of integrating the AC and RC training systems would be an increase in cross-component contact and cooperation. Having soldiers attend classes taught by instructors from other components, and possibly advancing to a point where AC and RC soldiers are trained in the same classroom, could strengthen confidence and understanding between the AC and the RC.

An additional benefit of sending AC soldiers to schools either collocated with their home locations or within an easy commute—including schools now "owned" by the RC—is a reduction in the time spent away from their home and unit. This would increase soldier morale.

RAND's Arroyo Center has been involved with TASS from its inception, starting with an assessment of the performance and efficiency of the prototype regional school system in the southeastern United States and subsequently providing recommendations for monitoring the efficiency and performance of the full system of RC schools as the TASS expanded to additional regions.

As part of that research, researchers determined that efficiencies could be gained within the RC school system by consolidating Annual Training (AT) training sites for courses with high support, supporting an effort by the Army National Guard (ARNG) to create regional sites for training in specific career fields.[2] Given the potential benefits discussed above, the next logical question was whether further efficiencies and economies of scale could be gained by integrating and leveraging resources across the Active and Reserve Components.

OBJECTIVES

This study extends the previous research to examine that question. More specifically, we developed a methodology for exploring the implications of shifting some training across existing component lines and used this method to illustrate potential benefits and tradeoffs of the approach. This report describes our methodology and presents exploratory analyses of three options for moving the TASS toward a more complete integration of AC and RC training infrastructure.

The **nearest school** option involves allowing students to attend the Army school closest to them, regardless of the component of the student or of the school. Under this option, RC soldiers will receive training at an AC proponent school if that school is closer than an RC school and AC soldiers will receive training at an RC school if it is

---

[2]For a more thorough discussion of this research, see Shanley et al. (1997).

closer than the AC proponent school, as long as the school is accredited and offers the course the student needs.

The **reassign courses** option entails modifying the courses offered at schools based on the localized demand for training. Here, we allow an RC school certified in a given functional area to offer AC and RC courses if there is sufficient demand within close proximity of the school. There are two cases within this option. In the *multifunctional* case, an RTS-M can offer a wide range of courses. In the *specialized* case, an RTS-M concentrates its course offerings in one or two functional areas.

The **consolidate schools** option considers the structure (i.e., the number of schools) needed to meet the integrated training requirement. Under this option, we examine the potential for existing RC schools to assume new training missions consistent with the localized demand for training.

SCOPE

It is important to note several issues about the exploratory analyses. First, they address only a subset of training: military occupational specialty (MOS) reclassification and functional training, and noncommissioned officer (NCO) education of enlisted personnel.[3] Currently, Initial Entry Training (IET) is conducted solely at AC training institutions under AC supervision (by law). Although our focus is on enlisted training, we believe there is also potential for integrating the training of officers.

Second, we selected a specific functional area as the focus of exploratory analyses—maintenance training conducted at the RC Regional Training Sites–Maintenance (RTS-Ms) and their AC proponent schools. These schools conduct over 100 courses in the maintenance area, including courses in enlisted leadership, MOS reclassi-

---

[3] NCO courses typically include a technical portion, specific to the MOS, and a general leadership portion. Completion of the Basic Noncommissioned Officer (BNCOC) course is required for promotion to the grade of E6. Completion of the Advanced Noncommissioned Officer (ANCOC) course is required for promotion to the grade of E7. Many MOS career fields end in a specific advanced leadership course; in some cases, several MOSs will feed into a single advanced course.

fication and advanced skills, and various sustainment and new equipment transition areas. In fiscal year 1996, almost 6,000 AC soldiers and over 8,000 RC soldiers received training in these types of courses (within their separate systems).

The 16 RTS-Ms in the continental United States plus the one in Hawaii provide an extensive network of alternative locations for RC maintenance training.[4] Although Aberdeen conducts the majority of the AC maintenance courses, Fort Lee, Fort Knox, Fort Jackson, Fort Leonard Wood, and Fort Sill also offer maintenance-related courses for AC soldiers.

Several of the RTS-Ms are collocated on Active bases (e.g., Fort Stewart, Fort Bragg, and Fort Hood) or are within an easy commute of one (e.g., Salina, Kansas is very close to Fort Riley). AC commanders have recognized the advantage of the collocated RC training facilities and have utilized the RTS-Ms to train their Active soldiers in various functional areas. In fiscal year 1996, over 1,000 AC soldiers took sustainment and modernization training courses at the local RTS-Ms.

The RTS-Ms in many ways are smaller versions of the AC schools. They have fixed facilities, permanently assigned training equipment, and a staff of full-time Active Guard and Reserve (AGR) administrators and instructors. They have close relationships with their AC proponent schools and have achieved accreditation and instructor certification. In fact, the instructors typically have greater teaching experience than their AC counterparts, with many having served at their RTS-Ms for several years.

Third, the three options examined only changes in the *location* of training. That is, we allow AC soldiers to take AC-configured courses at RC schools and RC soldiers to take RC-configured courses at AC schools. However, we do not commingle students from different components within the same classes, although we do allow the instructors of the classes to be from the other component (i.e., AC instructors could teach RC courses to RC students and RC instructors

---

[4]There are two other RTS-Ms, one at Tobyhanna, Pennsylvania, and one at Sacramento, California, that teach signal-related courses. Since we were focusing on maintenance courses, we have excluded these two from our analyses.

could teach AC courses to AC students). Of course, instructors from one component could be assigned, perhaps temporarily, to a school of the other component to assist in the training provided to any soldier.

The segregation of students could change, however, as Total Army Training System (TATS) courseware becomes available. Under TATS, one courseware package will be valid for all components. In principle, any student could take a course at any certified training location, taught by any qualified instructor, regardless of the component of the student, school, or instructor.

At this time, however, we are dealing with a relatively modest change, in which AC and RC courses are taught to AC and RC soldiers, respectively, albeit at any accredited Army school and by any certified instructor.

ORGANIZATION OF THIS DOCUMENT

In Chapter Two, we discuss the optimization modeling approach developed to explore the potential benefits of the three different options for integrating resources across the two components. Chapter Three discusses the results of applying that approach to the three options described above. Chapter Four presents some implications of the exploratory analyses.

Appendix A provides a technical description of the optimization model, and Appendixes B and C contain spreadsheets (from which the tables and figures in this document are derived) detailing the changes in student loads and course offerings for the various options we examined.

Chapter Two

DEVELOPING AN OPTIMIZATION MODEL FOR ANALYZING INTEGRATION OPTIONS

We built an optimization model to help analyze the various options. Below we briefly discuss what that model does in each of the three options we considered, the database used for the modeling, and the model's underlying assumptions.

HOW THE OPTIMIZATION MODEL WORKS

We developed a linear programming model designed to determine the least-cost assignment of students to schools under a variety of options by minimizing a subset of the total cost of providing training.[1] The costs included in the objective function are altered for the different options we examined. In the first option (termed "nearest school"), we changed only the *location* where a student receives training, holding fixed (at fiscal year 1996 levels) the schools and the courses offered at each school. Thus, in this option, the model minimizes the total travel cost, which is valued at 30 cents a mile.

In the second option (termed "reassign courses"), we allow *courses* to be "optimally" assigned to schools. In this case, the model's objective function includes both travel cost and annual fixed course cost.[2]

---

[1] A more thorough technical description of the model is provided in Appendix A.

[2] We break course costs into a fixed component, independent of the number of students taking the course at a specific school, and a variable component based on the student load. We specifically consider only the fixed component, assuming the variable component is similar across all AC and RC schools. Since we force all students to be assigned to a specific school, there is no change in the total variable course costs across all schools.

The annual fixed course cost—which we estimate at $50,000 based on our earlier research[3]—accounts for any instructor training or other "startup" costs necessary for a school to conduct a course. This fixed cost to offer a course drives the solution from multiple training locations with small class sizes (multifunctional) to a fewer number of locations with greater numbers of students (specialized).

In the third option (termed "consolidate schools"), we examined the *number of schools* that offer maintenance-related training. In this case, we add to the objective function a fixed cost for having a school open—$370,000, again drawn from data collected in our prototype assessment[4]—which is, in turn, added to the travel and fixed course costs. Similar to course costs, we only consider the annual fixed costs of having a school open. This "open the door" cost includes administrative staff personnel, utilities, and annual facility maintenance costs.

Given the different objective functions in the three options, the model minimizes the training costs while ensuring that every soldier who was trained in fiscal year 1996 is assigned to a training location (which may be different from the school he or she attended in fiscal year 1996) and that the student throughputs—the maximum annual number of students who can be assigned—of the school are not exceeded. The throughput for a school is based on its complement of facilities, instructors, and equipment.

Finally, the model has constraints related to the minimum number of students assigned to a school in each course. As we discuss in the next section, we looked at two situations. In the first case, we allow "multifunctional" schools that can teach a wide range of courses within their branch of certification. We require at least five students to offer a course. As would be expected, this case results in a number of locations teaching courses with a small number of students.

In the second case, we create "specialized" schools that offer a limited number of courses within their branch of certification. Here, we

---

[3] See Shanley et al. (1997) for a discussion of how this $50,000 cost was derived.

[4] See Shanley et al. (1997) for a discussion of how this $370,000 cost was derived. Similar to course costs, we only consider the annual fixed costs of having a school open.

add a constraint that requires at least 50 students in a grouping of related courses (e.g., wheeled or track courses) before a school will offer courses in that area. Here, courses are offered at fewer locations and schools offer a narrower range of courses.

MODEL DATABASE

In conducting the analyses, we consider all the AC and RC students who received training in fiscal year 1996 in NCO Education System (NCOES), MOS reclassification, and other maintenance-related courses (excluding initial entry training (IET) and advanced individual training (AIT)). Table 2.1 shows the number of AC and RC courses in the various areas that we analyzed.

Records from the Army Training Requirements and Resources System (ATRRS) show approximately 5,800 AC soldiers and 8,200 RC soldiers in this data set. A number of these records were missing data (primarily, the student's home origin) we needed for our analyses and, thus, had to be removed. We were left with 3,468 AC and 6,814 RC soldier records for our analysis.

MODEL ASSUMPTIONS

In using the model to conduct the analysis, we made two basic assumptions. First, as mentioned previously, we consider AC and RC courses as they were configured in fiscal year 1996. Thus, we assume an AC school can teach an RC version of any AC course that it offered

Table 2.1

Number of Maintenance-Related Courses

Type of Course	AC	RC
MOS reclassification	23	27
ANCOC	2	10
BNCOC	15	12
ASI	5	4
Other[a]	0	11

[a]Other includes sustainment, functional, and new equipment training.

in fiscal year 1996 and that RTS-Ms can teach AC versions of any RC course they offered in fiscal year 1996. For example, if an RTS-M offered an RC MOS reclassification course for 63B, we assume it could teach an AC version of the course.

The second assumption really deals with the type of model we are using. We are considering only the assignment of students to schools and not the scheduling aspects of the problem. Therefore, we make the assumption that a student who attended a course in fiscal year 1996 at a certain time and location could attend the course at any other time and location. This assumption would have little impact on AC soldiers who may be reassigned to other training locations, since they can attend training at any time. But it might have more of an effect on RC soldiers, although this could be mitigated if courses were offered on a regular basis at the nearby AC school, where most of these soldiers would be reassigned.

The objective of these exploratory analyses is to understand the feasibility and potential benefits of further integrating AC and RC training resources. As such, our initial assumptions are simplified and not necessarily complete. For actual policy use, the analytic model would need to be enhanced and the factors estimated in more detail.

Chapter Three

RESULTS OF ANALYZING THREE INTEGRATION OPTIONS

In this chapter we present the analysis results for our three options.

OPTION 1: NEAREST SCHOOL
Send Student to Nearest School Offering Like Course, Regardless of Component of School or Student

The first option examines the impact of sending AC and RC students to the nearest school that offers the maintenance course they need (MOS reclassification, NCOES, ASI, or other maintenance-related courses). We use the schools and their course offerings as reflected by the fiscal year 1996 ATRRS data. In this option, an RTS-M can offer an AC version of any RC course it offered in fiscal year 1996, and an AC school can offer an RC version of any AC course it offered.[1] We do not commingle students: AC students are in AC-configured courses and RC students are in RC-configured courses. Here, all we seek to do is minimize travel costs.

---

[1] In all options, we require a minimum of five students for a school to conduct a course. If the annual demand for a specific course is less than five students, then our minimum course size equals the annual demand. That is, courses with very low annual demands (less than ten) will be taught at a single school.

Sending Students to Nearest School Results in Significant Flows Across Components

One fairly straightforward result is that sending students to the nearest school results in significant flows across the two components. This is illustrated in Figure 3.1.

The chart on the left shows that 912 RC students (or about 13 percent of the 6,814 RC student records we processed) would go to an AC proponent school instead of an RTS-M. Also, 2,610 AC students (or about 75 percent of the 3,468 AC records we processed) would go to an RTS-M that is closer to their home location instead of to the AC proponent school.

The chart on the right shows the number of RC-configured courses at AC schools and the number of AC-configured courses offered at RTS-Ms. Note that the more than 120 AC courses at RTS-Ms do not represent 120 *different* courses; multiple RTS-Ms may be offering the same AC-configured course. With 17 RTS-Ms, there are on average only about 7 "new" AC-configured courses at each RTS-M. Almost one of every three students at an RT-M would be from the AC, while

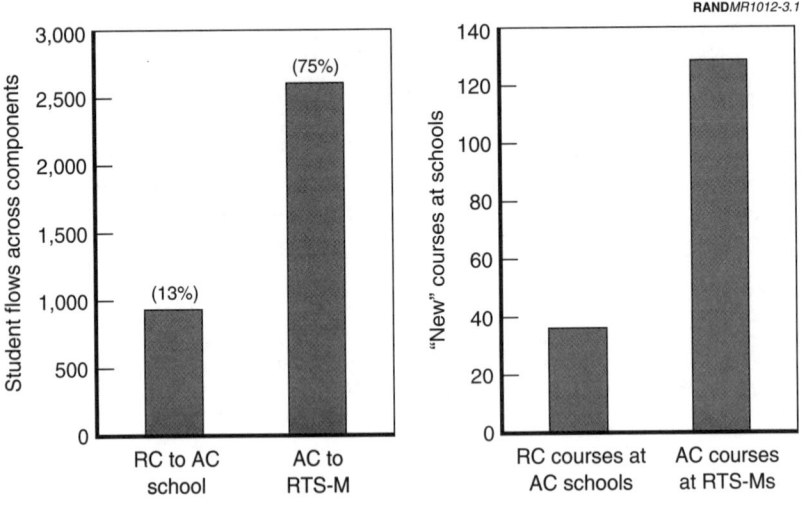

Figure 3.1—Flows for Sending Students in Option 1

slightly more than half the students at AC schools in the courses we are considering would be from the RC.

Table 3.1 shows the student flows by individual schools (10 AC schools on the top and 17 RTS-Ms on the bottom) for option 1, illustrating how the optimization model distributes students.[2]

The first two columns indicate the number of AC and RC students who went to the respective schools during fiscal year 1996 (the baseline case). In this case, as discussed above, there were 3,468 AC students in AC schools and 6,814 RC students in RC schools. The next two columns show the distribution of those AC and RC students when we allow a student to go to the nearest school offering the needed course regardless of the component of the school. The gray arrows show the AC student flows to AC and RC schools, while the black arrows show the RC student flows to RC and AC schools. We can see the breakdown of the 13 percent of RC students sent to AC schools and the 75 percent of AC students that end up in RTS-Ms. The last column shows the difference in the number of students at each school in our first option when compared to the baseline.

The table shows that several AC schools have their student loads significantly reduced or trade significant numbers of AC students for RC students. For example, there is a major impact at Fort Jackson USATC (most of the students in the 63B and 63S MOS reclassification courses and the 63B/S ASI course migrate to RTS-Ms); at Fort Leonard Wood USATC (the 66 AC students in the one 62B reclassification course migrate to RTS-Ms); at Fort Lee (most of the 231 students in the 92A course migrate to RTS-Ms); and at the NCO Academies (almost 1,400 AC NCOES students go to nearby RTS-Ms for training).[3]

---

[2]The detailed results, showing the specific courses taught at each school and the number of students across all the options, are provided in Appendix B.

[3]The parameters in the model can be set so that specific courses, e.g., NCOES, are fully excluded from the integration of AC and RC students. If the AC NCOES remain at the AC schools, the impact of integrating the MOS reclassification, ASI, and other courses is cut approximately in half. It might also be possible to preserve NCO leadership education at the proponent NCO Academies while migrating technical NCO education to nearby RTS-Ms. In this case, the impact of integration may be maintained.

Table 3.1
Specific AC and RC Student Flows in Option 1

	Baseline		Nearest school		
	AC	RC	AC	RC	Difference
AC schools					
Aberdeen	666	0	309	481	124
Fort Jackson USATC	497	0	25	25	(347)
Fort Knox	301	0	183	117	(1)
Fort L. Wood USATC	66	0	0	37	(29)
Fort Lee	231	0	28	10	(193)
Fort Sill	14	0	14	31	31
Aberdeen NCO Academy	1,201	0	287	58	(856)
Fort Knox NCO Academy	60	0	12	23	(25)
Fort L. Wood NCO Academy	45	0	0	3	(42)
Fort Lee NCO Academy	387	0	0	127	(260)
Total	3,468	0	858	912	(1,698)
			(25%)	(13%)	
RC schools (RTS-Ms)					
Fort Devens	0	68	8	22	(38)
Jefferson City	0	355	166	469	280
Fort Hood	0	402	239	373	210
Fort Indiantown Gap	0	242	74	309	141
Fort McCoy	0	252	291	206	245
Salina	0	573	134	412	(27)
Camp Dodge	0	963	109	480	(374)
Fort Dix	0	493	74	493	74
Fort Bragg	0	388	245	395	252
Camp Shelby	0	398	138	361	101
Camp Robert	0	578	233	566	221
Camp Ripley	0	381	85	333	37
Fort Custer	0	268	24	290	46
Gowen Field	0	408	182	314	88
Blanding	0	235	71	234	70
Fort Stewart	0	712	349	555	192
Waiwa	0	98	188	90	180
Total	0	6,814	2,610	5,902	1,698
			(75%)	(87%)	

Some of the RTS-Ms also get more students (e.g., Jefferson City, Fort Hood, Fort Bragg, and Camp Robert), while others have a decrease in

Figure 3.2—Effect on Training Workload at AC Schools and RC Schools

Figures 3.9 and 3.10 show the impacts on travel cost and time away from home, respectively, for the specialized case of option 3. As in the figures for comparable scenarios of option 1 and option 2, the travel cost savings for option 3, as compared to those of option 1, of more than $1,000,000 savings in AC student travel is offset by a $500,000 increase in RC student travel.

Finally, Table 3.8 shows for selected courses the number of sites offering courses in the specialized case of option 3.

Number of PTS-Ms Can Be Reduced

The model allocates the fiscal year 1996 training inputs more than the 12 RTS-Ms currently conducting BT maintenance training. As shown above in Table 3.7, two PTS-Ms in our specialized case are "excess" for the maintenance training demand. We do not suggest that these "excess" or standby facilities could be immediately disposed to facilitate executing equipment and redirecting course/unit training missions to other installations.

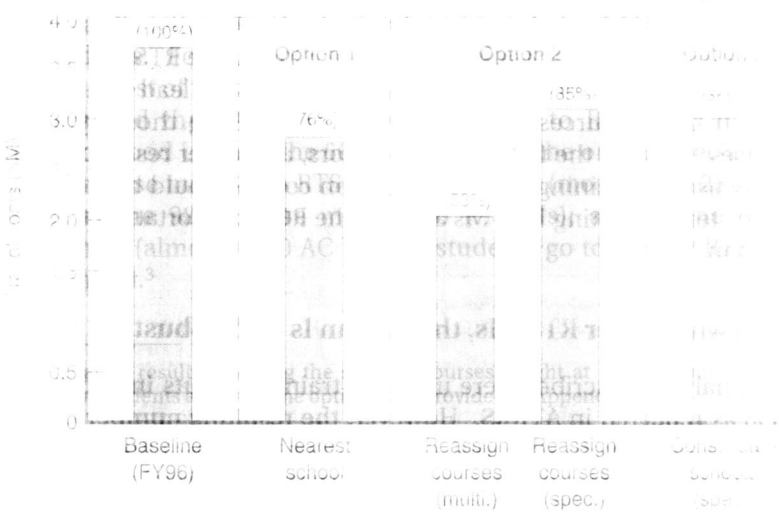

Figure 3.9—Impact of Travel Cost for Option 3 Compared to Options 1 and 2

Table 3.5

Specific AC and RC Student Flows in Option 2 (Specialized Case)

	Baseline		Reassign courses (specialized)		
	AC	RC	AC	RC	Others
Schools					
Aberdeen	666	0	242	403	21
Jackson USATC	497	0	136	195	160
BLISS	301	0	183		118
Leonard USATC	66	0	0		66
Sill	231	0	190	0	41
	14		0		
Infantry School Academy	1,201	0	184		911
Cavalry Academy	60	0	0		
NCO Academy	45	0	0		
NCO Academy	387	0	164	136	
Total	3,468	0	1,009	911	
			(29%)	(13%)	
RC ARNG					
	0	88	12	0	87
	0	355	213	153	60
	0	402	292	258	48
	0		36	42	85
	0	252	96	161	
	0	175	144	412	14
	0	302	165	411	357
	0		41	300	106
	0	260	210	353	106
	0	392	188	278	60
	0	370	197	180	82
	0	355	189	195	
	0	268	107	415	152
	0	408	120	337	369
	0		206	312	
	0		129	50	108
	0		63		
Total	0		2,459	3,896	
			(71%)	(87%)	

Results of Analyzing Three Integration Options   23

Table 3.5 shows the counterpart to Table 3.1 for option 1, in this case for the specialized case of option 2. As was true in option 1, the AC schools, especially the NCO Academies, trade a large number of AC students for a smaller number of RC students. In the specialized case, the AC schools with small student loads—Fort Leonard Wood USATC and the NCO Academy, Fort Sill, and Fort Knox NCO Academy—have all their AC maintenance students reassigned to RTS-Ms. Also, as with option 1, some RTS-Ms have more students while others have fewer or stay about the same.

Travel Costs Decrease in All Options

Figure 3.6 shows the impact on travel costs. There is a separate bar for the baseline calculated from the actual assignment of students to schools in fiscal year 1996 and bars for the first option (sending students to the nearest school based on the fiscal year 1996 course assignments) and the two cases of option 2. In all options, travel costs are reduced compared to the baseline, with the greatest reduction, as expected, when we allow RTS-Ms to teach multifunctional courses. Table 3.6 shows that over $1,000,000 is saved in AC student travel and almost $500,000 is saved in RC student travel for the multifunctional case. The specialized case shows different results. The more than $1,000,000 savings in AC student travel is offset by a $500,000 increase in RC student travel. The net effect for the specialized case is a decrease in total travel costs of approximately $500,000.

AC Student Time Away from Home Also Decreases in All Options

Finally, Figure 3.7 shows the impact on AC students' time away from home. Again, there is a bar for the baseline where the AC student goes to the AC school for training and education and separate bars for the two options that allow students to take courses regardless of the component of the student or of the school. In all the options, the time away from home is reduced. This has the impact of improving morale or quality of life while reducing per-diem costs.

Table 3.5
Specific AC and RC Student Flows in Option 2 (Specialized Case)

	Baseline		Reassign courses (specialized)		
	AC	RC	AC	RC	Difference
AC schools					
Aberdeen	666	0	242	403	(21)
Fort Jackson USATC	497	0	136	195	(160)
Fort Knox	301	0	183	0	(118)
Fort L. Wood USATC	66	0	0	52	(14)
Fort Lee	231	0	100	0	(131)
Fort Sill	14	0	0	33	19
Aberdeen NCO Academy	1,201	0	184	76	(941)
Fort Knox NCO Academy	60	0	0	23	(37)
Fort L. Wood NCO Academy	45	0	0	0	(45)
Fort Lee NCO Academy	387	0	164	136	(87)
Total	3,468	0	1,009	918	(1,541)
			(29%)	(13%)	
RC schools (RTS-Ms)					
Fort Devens	0	68	13	0	(55)
Jefferson City	0	355	213	234	92
Fort Hood	0	402	292	258	148
Fort Indiantown Gap	0	242	86	429	273
Fort McCoy	0	252	96	161	5
Salina	0	573	144	415	(14)
Camp Dodge	0	963	165	441	(357)
Fort Dix	0	493	41	598	146
Fort Bragg	0	388	210	334	156
Camp Shelby	0	398	188	278	68
Camp Robert	0	578	197	463	82
Camp Ripley	0	381	189	389	197
Fort Custer	0	268	107	313	152
Gowen Field	0	408	120	657	369
Blanding	0	235	206	312	283
Fort Stewart	0	712	129	567	(16)
Waiwa	0	98	63	47	12
Total	0	6,814	2,459	5,896	1,541
			(71%)	(87%)	

Results of Analyzing Three Integration Options   25

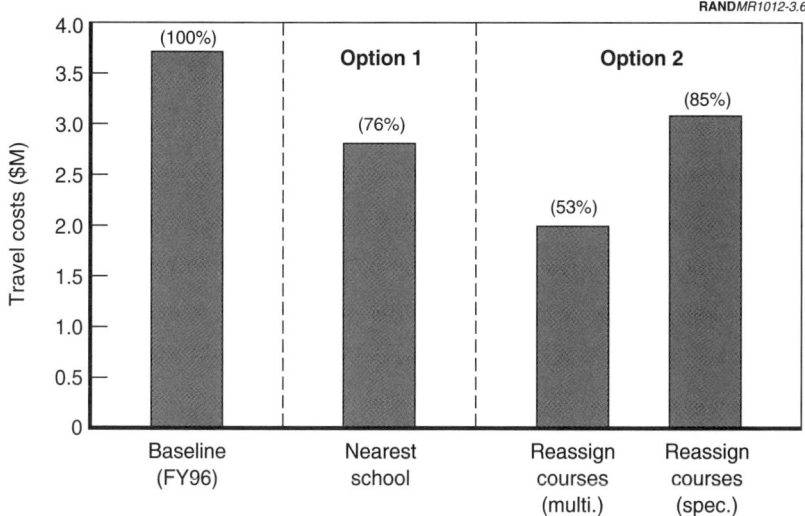

Figure 3.6—Impact of Travel Costs for Option 2 Compared to Option 1

Table 3.6

Breakdown of Travel Cost Savings for Option 2

Cost of Sending...	Baseline	Reassign Courses (Multifunctional)	Reassign Courses (Specialized)
AC students to AC schools	2,099,182	290,580	301,852
AC students to RTS-Ms		592,727	731,739
RC students to RTS-Ms	1,569,044	923,137	1,588,997
RC students to AC schools		146,059	476,935

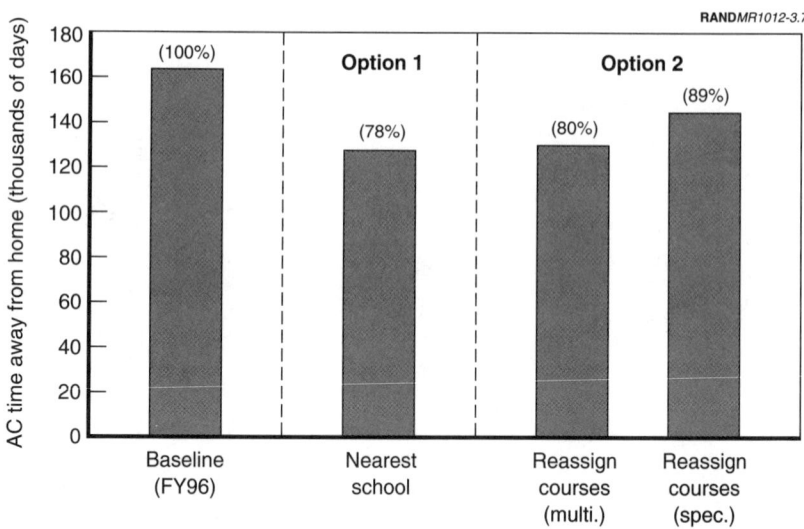

Figure 3.7—AC Student Time Away from Home for Option 2 Compared to Option 1

OPTION 3: CONSOLIDATE SCHOOLS
Reduce Number of Training Sites for Maintenance Courses

The last option we examined investigates the potential for reducing the number of training sites for maintenance courses. Depending on the specific case examined (multifunctional or specialized RTS-Ms), the model indicates that maintenance courses could be offered at from 2 to 6 fewer RTS-Ms than the 17 currently used.[7] Although these RTS-Ms could potentially be closed, we believe that a greater benefit would arise from changing the training mission of these "excess" schools.

---

[7]Since the Army is moving away from multifunctional schools and since it is unlikely to make reductions on the scale that the model suggests, we have focused the analysis on the specialized case.

We add to the model's objective function a fixed cost of $370,000 to open an RTS-M. Therefore, the model balances savings in travel and course costs with the fixed costs of an RTS-M.

Results Mirror Those for Previous Options

The results from this option mirror the results from the previous two options. A significant number of students and courses cross component boundaries, with a resulting decrease in travel costs plus AC student time away from home and per-diem costs. Figure 3.8 shows the student flows for all three options, using the "specialized" case for option 3. As can be seen, the 10 percent–70 percent numbers are about the same as in the other two options.

Table 3.7 shows the specific student flows for the specialized case in option 3. In this case, the model suggests that two RTS-Ms could be closed or have their mission changed.

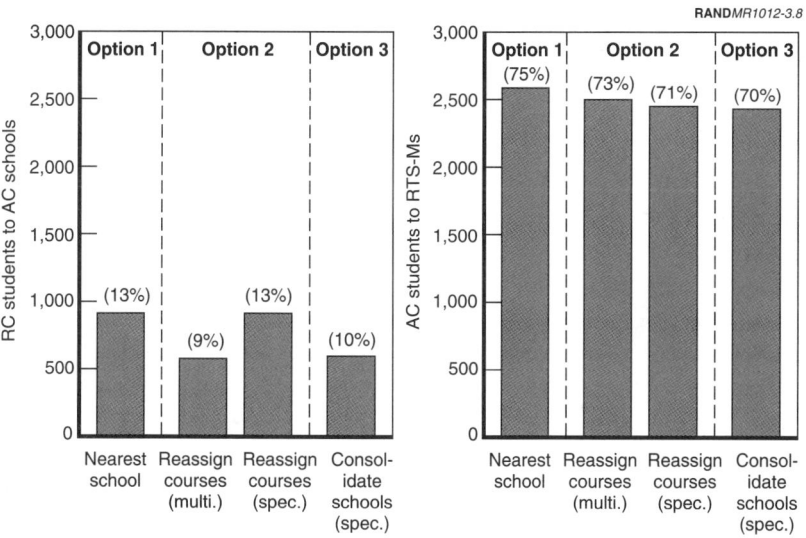

Figure 3.8—Student Flows of RC Students to AC Schools and AC Students to RTS-Ms for Option 3 Compared to Options 1 and 2

Table 3.7

Specific AC and RC Student Flows in Option 3 (Specialized Case)

	Baseline		Consolidate schools (specialized)		
	AC	RC	AC	RC	Difference
AC schools					
Aberdeen	666	0	238	385	(43)
Fort Jackson USATC	497	0	157	50	(290)
Fort Knox	301	0	175	0	(126)
Fort L. Wood USATC	66	0	0	51	(15)
Fort Lee	231	0	100	0	(131)
Fort Sill	14	0	0	33	19
Aberdeen NCO Academy	1,201	0	193	36	(972)
Fort Knox NCO Academy	60	0	0	23	(37)
Fort L. Wood NCO Academy	45	0	0	0	(45)
Fort Lee NCO Academy	387	0	180	136	(71)
Total	3,468	0	1,043	714	(1,711)
			(30%)	(10%)	
RC schools (RTS-Ms)					
Fort Devens	0	68	13	0	(68)
Jefferson City	0	355	221	381	247
Fort Hood	0	402	290	195	83
Fort Indiantown Gap	0	242	149	464	371
Fort McCoy	0	252	0	0	(252)
Salina	0	573	159	345	(69)
Camp Dodge	0	963	170	297	(496)
Fort Dix	0	493	30	659	196
Fort Bragg	0	388	161	507	280
Camp Shelby	0	398	198	301	101
Camp Robert	0	578	202	451	75
Camp Ripley	0	381	156	499	274
Fort Custer	0	268	167	313	212
Gowen Field	0	408	133	698	423
Blanding	0	235	191	377	333
Fort Stewart	0	712	136	557	(19)
Waiwa	0	98	62	56	20
Total	0	6,814	2,425	6,100	1,711
			(70%)	(90%)	

Figures 3.9 and 3.10 show the impacts on travel cost and time away from home, respectively, for the specialized case of option 3. Again, the figures are comparable across all the options. The breakout of travel cost savings for option 3 are similar to those of option 2—the more than $1,000,000 savings in AC student travel is offset by a $500,000 increase in RC student travel.

Finally, Table 3.8 shows for selected courses the number of schools offering courses in the specialized case of option 3.

Number of RTS-Ms Can Be Reduced

The model allocates the fiscal year 1996 training inputs into fewer than the 17 RTS-Ms currently conducting RC maintenance courses: As shown above in Table 3.5, two RTS-Ms in the specialized case are "excess" for the maintenance training demand. We do not, however, suggest that these "excess" training facilities be closed. Since they have very capable facilities, training equipment, and personnel, redirecting their training mission to other nonmaintenance areas,

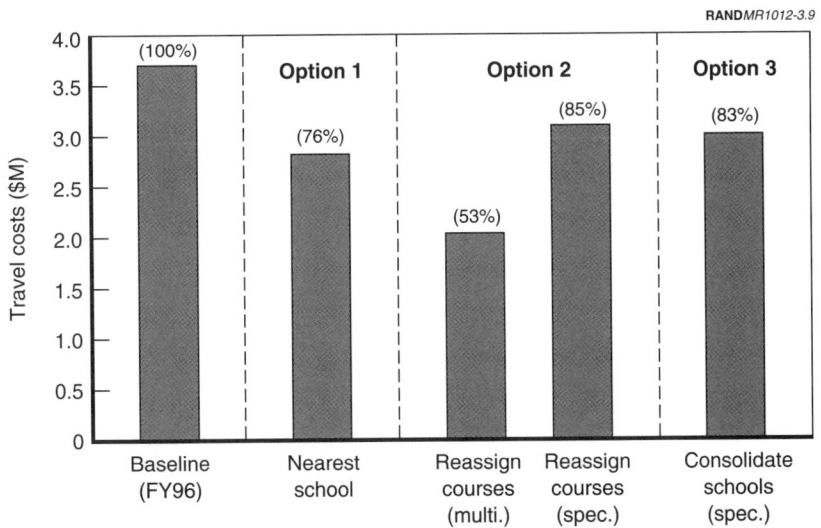

Figure 3.9—Impact of Travel Cost for Option 3 Compared to Options 1 and 2

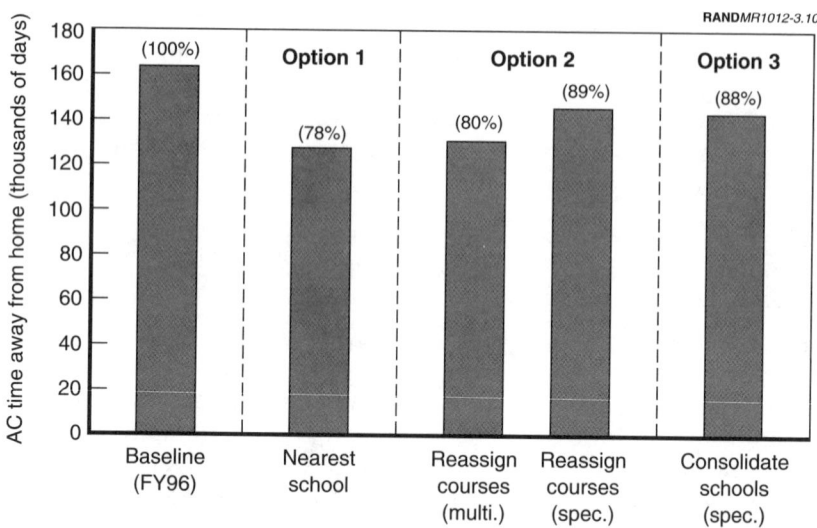

Figure 3.10—AC Student Time Away from Home for Option 3 Compared to Options 1 and 2

such as transportation or quartermaster courses, might be a more cost-effective use of those sites. Of course, using the RTS-Ms to provide training in areas such as transportation could lead to a reduction in the resources currently used for teaching those types of courses. That is, the facilities, instructors, and other resources currently used in training RC transportation courses could be excess to the system if existing RTS-Ms assume the RC transportation training mission.

Even with Fewer RTS-Ms, the System Is Still Robust

The analyses described here used the training inputs in fiscal year 1996 as reflected in ATRRS. However, the resulting number of students, especially RC students, may not represent the actual number of soldiers that required maintenance training in fiscal year 1996.[8]

---

[8] Our earlier research showed that the number of RC soldiers shown as not duty MOS qualified greatly exceeded the number of seats available in RC schools. Similarly, we

Table 3.8
Number of Schools Offering Selected Courses: Option 3

Course ID	Component	Course Level	MOS	FY96 Students	Baseline	Consolidate Schools (Spec.)
091-52D10	RC	MOSQ	52D	231	12	5
052-62B10	RC	MOSQ	62B	201	12	4
052-62B30	RC	BNCOC	62B	65	9	2
091-63B10	RC	MOSQ	63B	723	17	14
091-63B30	RC	BNCOC	63B	298	15	9
091-63-B/S/W10H8	RC	ASI	63B/S/W	141	11	5
551-92A10	AC	MOSQ	92A	231	1	5
612-62B10	AC	MOSQ	62B	66	1	4
612-62B30	AC	BNCOC	62B	45	1	2
610-63B10	AC	MOSQ	63B	159	1	6
610-63B30	AC	BNCOC	63B	331	1	11
610-ASIH8 (63B/S)	AC	ASI	63B/S	286	1	6

Also, there may be future surges in the demand for maintenance-related training because of force structure changes or turbulence in the RC personnel system. Any changes to the structure and use of schools providing maintenance training must ensure that the resulting structure is robust enough to meet demands above those reflected by the actual training inputs in fiscal year 1996.

To examine how robust the system would be with up to six fewer RTS-Ms, we used Standard Installation/Division Personnel System (SIDPERS) data to estimate the number of RC soldiers who are shown as nonqualified and who, hence, required maintenance-related training in fiscal year 1996. We checked individual soldier records to see which soldiers were not duty MOS qualified (DMOSQ) or required NCOES training in fiscal year 1996. The result may be an overestimate of requirements (since some soldiers will either shortly

---

observed a "backlog" of NCOs who needed to complete the NCOES course required for current or impending grade; this number greatly exceeded available classroom seats. In general, the number of available seats was less than half the number of soldiers showing a need for reclassification training or NCOES. (See Winkler et al., 1996.)

leave the force or will transition to a new MOS), but if the school structure can accommodate an overestimate of requirements, then it should surely be able to handle the "true" demand.

To explore whether these schools could handle an expanded demand for training, we examined whether the reduced RTS-M structure could handle a demand that was approximately twice the load shown in ATRRS in fiscal year 1996. We found that the reduced RTS-M structure still had the capacity to accommodate the increased training load (assuming the remaining RTS-Ms had their full TDA complement of instructors). The total number of AC and RC instructors needed to meet the training workload indicated by the fiscal year 1996 inputs is sufficient for the increased demand, although there may be some redistribution of instructors among the schools, both by type (especially for the "specialized" RTS-Ms) and in number. For example, with training workload redistributed between the AC and RC schools, some AC instructors might be assigned temporarily to an RTS-M to help meet peak demands.

RTS-Ms Have the Capacity to Assume New Missions

To examine the potential for the "excess" RTS-M capacity to take on other training missions, we extracted from ATRRS the number of AC and RC soldiers who received training in transportation-related courses in fiscal year 1996. This was approximately 3,400 RC and 800 AC soldiers. The model suggests that the RTS-Ms have sufficient capacity to meet this demand for transportation courses in addition to the increased maintenance-related demand described above. Rather than mixing maintenance and transportation courses at RTS-Ms, we believe it would be more effective to have some number of the RTS-Ms concentrate on transportation courses (as RTS-Ts), while the remaining schools conduct the maintenance courses.

Chapter Four
CONCLUSIONS AND IMPLICATIONS

Although the analyses conducted here were exploratory, there are still some important implications from the research. Basically, the implications have to do with the benefits of further integration, whether the maintenance-related results can be generalized to other areas, and the potential next steps for the Army.

AC/RC SCHOOL INTEGRATION PROVIDES A RANGE OF PERSONNEL-RELATED BENEFITS

The exploratory analysis suggests that there are benefits for three personnel groups: instructors, school support staff, and AC and RC students. We examine each in turn.

Instructor Benefits

Our analyses suggest that allowing students to attend the nearest school offering a course, regardless of the component of the student or the school, can result in a large number of student training days migrating from the AC schools to the RTS-Ms. This migration reduces the platform time and workload of AC instructors, thereby providing a boost in morale to what may be a currently overworked and overstressed workforce. In addition, many believe that the current cadre of RC instructors can accommodate this increased workload, particularly given current levels of quota utilization.

Although fewer overall instructors may be needed in the integrated system, we stress that a more thorough analysis is needed to understand how best to utilize the AC, AGR, and part-time RC instructors. For example, sending an AC instructor temporarily to schools close to the students may be preferred to sending the students to a school some distance from the location of their home. Moreover, temporarily assigning RC instructors, either full-time AGRs or part-time M-day personnel, to AC schools may be an efficient way to satisfy peak demands.

Support Personnel Benefits

In terms of support personnel, we found that the number of school support personnel is fairly insensitive to the training workloads at the school. Therefore, increasing the workload at the RTS-Ms, or decreasing it at the AC schools, should have almost no effect on the number of school support personnel. If RTS-Ms are closed, however, or have their missions changed, there may be some support staff personnel savings in the system.

Student Benefits

The greatest advantage of furthering the integration of the AC and RC school systems may be associated with the students. Travel time and cost can be reduced, and, most important, the time away from home (and the resulting per-diem costs) will be reduced for AC soldiers who go to an RTS-M either at their home base or within an easy commute. Also, the AC does not have separate courses for MOS reclassification. AC soldiers needing training in a new MOS go to the same course as soldiers receiving their first MOS training. A portion of these "full" MOS courses addresses common skills and provides soldierization beyond what is needed for a prior-service soldier. RC MOS reclassification courses assume that the soldiers have received the common skill and soldierization portions during IET and, therefore, are shorter than the corresponding initial entry course. If AC soldiers needing reclassification training in a new MOS were permitted to take the RC version of the course, overall AC training time could be reduced.

Although many of the advantages are monetary, integration of the training and education of AC and RC soldiers would presumably enhance the overall integration of the total force. When students are taught by instructors from other components, in the facilities of the other components, and potentially alongside students from the other components, one could expect that cultural barriers would begin to disappear and cross-component contact and confidence would increase.

RESULTS CAN BE GENERALIZED BEYOND MAINTENANCE TRAINING AREA

Our analysis has focused on maintenance training. However, we believe the concepts and opportunities for further integrating AC and RC training exist in other areas as well. The most obvious extensions beyond maintenance would be to other functional areas with regional training sites in the RC (e.g., engineer, medical, and military intelligence). However, we think that in principle the concept could be extended readily to those RC training organizations that operate regional training sites, such as maintenance, medical, and intelligence.

The alternative training opportunities and the potential for more efficient operations largely result from the RTS facilities and personnel that may already exist. RTSs are miniversions of their AC counterparts with fixed facilities, permanent training equipment, and full-time personnel. The training of RC soldiers could greatly benefit from similar structures in other areas, such as the conversion of some existing RTS-Ms to other functions (e.g., to Regional Training Sites for Transportation (RTS-Ts)) or the development of similar regional RC schools in other functional areas.

PILOT-TESTING AC/RC INTEGRATION AND CONDUCTING MORE DETAILED ANALYSES OF INSTRUCTOR REQUIREMENTS ARE LOGICAL NEXT STEPS

The foregoing analysis indicates that there are potential benefits from integrating training across components. If the Army should decide to proceed along these lines, the next step should be a pilot test, to better understand the options and the policy and resource implications. Such a pilot program might include selecting two or three RTS-Ms, potentially those collocated at active bases, to conduct AC-configured courses. Aberdeen, or another AC school, could begin to offer RC-configured MOS reclassification and possibly NCOES courses. One or more RTS-Ms, potentially those with low student workloads, could also offer transportation or quartermaster courses on a trial basis.

Admittedly, these changes could have significant implications if found beneficial and adopted on a wider scale. The integration of reclassification training in the maintenance area seems straightforward enough; however, it could result in a shorter reclassification course for AC soldiers, along the lines of (and perhaps in conjunction with) current RC reclassification training. The integration of NCOES could face strong cultural resistance, since some AC NCOs would then take some or all their BNCOC or ANCOC at an RTS-M (presumably with oversight from a proponent or major command NCO Academy). However, some RC NCOs will now attend AC NCO Academies. While such changes seem dramatic, it may be impossible to maintain the status quo. Reduced training budgets and TDA personnel allowances at the schools require the Army to find more efficient ways to conduct the training of all soldiers, regardless of component.

A second desirable step, as we have noted previously, would be a more thorough examination of the instructor requirements resulting from a more fully integrated school system. A potential barrier to integration arises from Army manpower staffing standards, which "reward" shifts in workload with reductions in manpower resources. An estimate of how many instructors, and what type of instructors (e.g., AC, AGR, or part-time RC), are needed at various training locations is necessary to better understand potential cost and resource implications and to identify options for encouraging such innovation. Such analyses should consider the potential for temporarily assigning instructors to training locations, either by detailing them from their assigned school to another school or by using part-time (i.e., M-day) RC instructors, to help meet a surge in training workload and encourage systemwide efficiency.

Appendix A
TECHNICAL DESCRIPTION OF THE OPTIMIZATION MODEL

This appendix provides a technical description of the optimization model developed to examine the options presented in the body of the report. The model uses a linear programming construct and is coded in the GAMS software package.

MODEL DIMENSIONS

There are five dimensions, or subscripts, used in the model:

- Subscript i represents the two Army components, AC and RC.
- Subscript j represents the 51 potential home locations (50 states plus the District of Columbia) of the students.
- Subscript k represents the 111 maintenance-related courses included in the analysis. A list of the courses is provided in Appendix C.
- Subscript l represents the 27 different AC (10 schools and NCO Academies) and RC (17 RTS-Ms) schools that conduct maintenance-related courses. A list of the schools is provided in Appendix B.
- Subscript g represents the 11 different course groupings. The relationship of courses to groups is based on department structure at Aberdeen or on the courses offered at the other AC schools. The MOSs or functional courses in each group are shown in Table A.1.

Table A.1
Relationship Between Groups and MOSs

Group Name	MOSs/Functional Courses
ANCOC	All AC 63 level
Construction equipment	62B
Field artillery	45D, 63D
Metalworking	44B, 44E
Quartermaster	92A, TAMMS, ULLS
Tactical support equipment	52C, 52D, 52F, 63J
Tanks and Bradleys	45E, 45T, 63E, 63N, 63T
TOW/Dragon repair	27E
Weapons	45B, 45G, 45K
Wheel and track vehicle (DS/GS)	63G, 63H, 63W, 63Y
Wheel vehicle (organizational)	63B, 63S

DECISION VARIABLES

There are four decision variables in the model:

- $X_{i,j,k,l}$ is the number of students in component i from home location j taking course k at school l.

- $W_{k,l}$ is a binary (0,1) variable indicating that course k is offered at school l. This variable can be preset in the model to force certain courses to be offered at certain schools (option 1), or the model can determine the optimal assignment of courses to schools (options 2 and 3).

- Y_l is a binary (0,1) variable indicating school l is open. This variable can be preset in the model to force a school to be open (options 1 and 2), or the model can determine the optimal set of schools to use for the course offerings (option 3).

- $T_{g,l}$ is a binary (0,1) variable indicating that course group g is offered at school l. This variable can be preset to indicate that a set of courses are offered at a school (option 1 and 2 multifunctional), or the model can determine the optimal assignment of course groupings to schools (option 2 specialized and option 3).

PARAMETERS

There are a number of parameters included in the model:

- $D_{j,l}$ is the round-trip distance in miles from a student's home location j to school l. Data from the ATRRS files provide the home state for RC students and the state where AC students' units are located. The model uses the latitude and longitude for the centroid of each state and the latitude and longitude for each school to calculate the straight-line distance.
- E_k is the length of course k in days. These data are from ATRRS.
- $\$D$ is the cost per mile. A factor of .30 is used.
- $\$F$ is the fixed annual cost for offering a course. This cost includes initial supplies and courseware and the required training for the course instructor. We use a value of $50,000 based on earlier research.[1]
- $\$_l$ is the fixed annual cost for having school l open. It includes administrative staff personnel, utilities, and annual facilities maintenance costs. We use a factor of $370,000 for all schools and RTS-Ms.
- $R_{i,j,k}$ is the number of students from component i at home location j requiring course k. These data are from the fiscal year 1996 ATRRS files and represent the actual number of students who were trained.
- M_k is the minimum number of students required to offer course k. We use a value of 5 students for all courses except those courses where the annual demand was less than 5 students. For those low-demand courses, we use the annual demand as the minimum class size.

---

[1] See Shanley et al. (1997) for a discussion of how this $50,000 cost was derived.

- M_g is the minimum number of students in group g in order for a school to offer courses in that group. We use a factor of 50 for all groups.
- $Q_{k,l}$ is the maximum number of students in course k at school l. These data are from the ATRRS schedule file.
- Q_l is the maximum number of student days at RTS-M l. We use a value of 15,600 for all RTS-Ms. This value is based on an average of data from the RTS-Ms at Forts Stewart, Bragg, and Dix. It considers the number of classrooms and instructors available at an RTS-M. It relates to 1,200 students a year taking an average 13-day course.
- $S_{k,g}$ is a binary variable set to one if course k is in group g, and zero otherwise.

MODEL OBJECTIVE FUNCTION

The objective of the model is to minimize costs. For the general model, the mathematical representation is

$$\text{Minimize } \sum_i \sum_j \sum_k \sum_l X_{i,j,k,l} (D_{j,i} \$D) + \sum_k \sum_l \$F\, W_{k,l} + \sum_l \$_l\, Y_l.$$

The first term represents the travel cost and the variable course costs (currently set to zero) associated with sending X students in component i from location j to school l to take course k. The second term captures the fixed cost of offering a course at a school. The last term represents the fixed cost in having a school open.

In option 1 and the multifunctional case of option 2, we use only the first cost element, the travel cost (i.e., we set $\$F$ and $\$_l$ equal to zero). For the specialized case of option 2, we use the first two elements of cost, the travel and fixed course costs (i.e., we set $\$_l$ equal to zero). For option 3, we use all three elements of cost.

MODEL CONSTRAINTS

There are several constraints in the model that define the potential solution space for the objective function. We list these constraints below and describe how they are, or are not, used for the specific options we analyzed.

Supply Equals Demand

$$\Sigma_l X_{i,j,k,l} = R_{i,j,k} \text{ for all } i, j, k.$$

In each of the options, we ensure that the AC and RC students who were trained in fiscal year 1996 in the maintenance-related courses we are considering are assigned to an AC school or an RTS-M.

Assign Students to Schools Where the Course Is Taught

$$\Sigma_i \Sigma_j X_{i,j,k,l} \leq \Sigma_i \Sigma_j R_{i,j,k} W_{k,l} \text{ for all } k, l.$$

In each of the options, students can only take courses at the AC schools or the RTS-Ms where the courses are offered. In option 1 and the multifunctional case of option 2, we use the fiscal year 1996 assignment of courses to schools as reflected in ATRRS to predefine the $W_{k,l}$ binary variable. For the specialized case in option 2 and for option 3, we allow the model to determine the least-cost assignment of courses to schools based on the local demand for a course.

Ensure Minimum Course Size Requirements Are Met

$$\Sigma_i \Sigma_j X_{i,j,k,l} \geq M_k W_{k,l} \text{ for all } k, l.$$

This constraint is used for all options. It prevents the model from assigning less than a minimum number of students to a school for a specific course.

Ensure Maximum Course Size Is Not Exceeded

$$\Sigma_i \Sigma_j X_{i,j,k,l} \leq Q_{k,l} \text{ for all } k, l.$$

This constraint is used for all options and prevents the model from assigning more students to a school for a specific course than the maximum number that a school can accommodate.

Ensure Maximum Capacity of a RTS-M Is Not Exceeded

$$\sum_i \sum_j \sum_k X_{i,j,k,l} E_k \leq Q_l \text{ for all RTS-Ms.}$$

This constraint is used for all options. It ensures that the capacity of an RTS-M, expressed in annual student days, is not exceeded. That is, the total number of student days associated with a number of students taking various courses must be within the capacity of the RTS-M.

Assign Courses to Schools Where Their Groups Are Taught

$$\sum_i \sum_j \sum_k X_{i,j,k,l} S_{k,g} \leq 10{,}000 \; T_{g,l} \text{ for all RTS-Ms } l \text{ and groups } g.$$

This constraint is used for the specialized case of option 2 and for option 3. It makes sure that courses can be taught at schools that have the responsibility for the group to which the course belongs. The 10,000 factor is an arbitrarily large number that ensures if $T_{g,l}$ is one, then courses in that group can be taught at the school.

Comply with the Minimum Number of Student Days Needed for a Course-Group to Be Taught at an RTS-M

$$\sum_i \sum_j \sum_k X_{i,j,k,l} S_{k,g} \geq 50 \; T_{g,l} \text{ for all RTS-Ms } l \text{ and groups } g.$$

This constraint is used for the specialized case of option 2 and for option 3. It ensures that groups are assigned to schools only when there are at least 50 students taking courses within that group.

Teach Courses Only at Schools That Are Open

$$W_{k,l} \leq Y_l \text{ for all } k \text{ and } l.$$

This constraint is used in option 3. It ensures that courses are offered only at schools that remain open (i.e., that are assigned maintenance-related courses).

Appendix B
COURSES TAUGHT AT EACH SCHOOL FOR THE VARIOUS OPTIONS

This appendix shows the specific courses offered at each of the AC schools and the RTS-Ms for the various options described in the body of the document. For each school, the tables show the course identifiers, the specific phases for the RC courses, the course level (MOS reclassification, Advance Skill Identifier (ASI), BNCOC, ANCOC, etc.), the MOS appropriate for the course, and the number of AC and RC students in each course and in total for the base case (i.e., as reflected in the fiscal year 1996 ATRRS database) and for each of the three options examined during the analyses.

Course ID	Phase	Course Level	MOS	School
052-62B10	(blank)	MOS	62b1	Ft. Devens RTSM
				Jefferson City RTSM
				RTS-M, FT HOOD, TX
				RTS-M, FT INDNTWN GP PA
				RTS-M, FT MCCOY, WI
				RTS-M-02 CAMP DODGE IA
				RTS-M-04 FT BRAGG NC
				RTS-M-05 CAMP SHELBY MS
				RTS-M-06 CAMP ROBERT CA
				RTS-M-07 CAMP RIPLEY MN
				RTS-M-08 FT CUSTER MI
				RTS-M-09 GOWEN FIELD ID
				USATC, FT. WOOD/98TH DIV
052-62B10 Total				
052-62B10 (T)	(blank)	Transition	62B	RTS-M-12 FT STEWART GA
				USATC, FT. WOOD/98TH DIV
052-62B10 (T) Total				
052-62B30	2	BNCOC	62b3	Ft. Devens RTSM
				Jefferson City RTSM
				NCO ACADEMY - FT L. WOOD
				RTS-M, FT HOOD, TX
				RTS-M, FT INDNTWN GP PA
				RTS-M-02 CAMP DODGE IA
				RTS-M-04 FT BRAGG NC
				RTS-M-05 CAMP SHELBY MS
				RTS-M-09 GOWEN FIELD ID
				RTS-M-12 FT STEWART GA
052-62B30 Total				
052-62B40	2	ANCOC	62B4	Jefferson City RTSM
				USATC, FT. WOOD/98TH DIV
052-62B40 Total				
091-44B10	1	MOS	44b1	ORDNANCE SCH, APG
				RTS-M-01 SALINA KS
				RTS-M-12 FT STEWART GA
	2	MOS	44b1	ORDNANCE SCH, APG
				RTS-M-01 SALINA KS
				RTS-M-12 FT STEWART GA
	3	MOS	44b1	ORDNANCE SCH, APG
				RTS-M-01 SALINA KS
				RTS-M-12 FT STEWART GA
091-44B10 Total				

Courses Taught at Each School for the Various Options

Base Case		Nearest School		Reassign Courses Multifunctional		Reassign Courses Specialized		Consolidate Schools Specialized	
AC	RC	AC	RC	AC	RC	AC	RC	AC	RC
	11		9		9				
	20		5						
	11		7		5		17		
	14		32		30		73		47
	9		12		10				
	36		6		5				
	47		36		30		50		44
	7		21		21				
	7		6		6				
	6		14		14				48
	11		11		23				
	22		21		21		27		27
			21		27		34		35
	201		201		201		201		201
	2		2						2
					2		2		
	2		2		2		2		2
	3		3		9				
	6		4						
			3		9				
	8		3		5				
	2		12		9		28		52
	14		7		6				
	6		7		5				
	10		7		5				
	3		10		10		13		13
	13		9		7		24		
	65		65		65		65		65
	16								
			16		16		16		16
	16		16		16		16		16
			45		7		45		45
	27				24				
	18				14				
					5				
	10		20		7		20		20
	10				8				
							9		9
	6				9				
	3		9						
	74		74		74		74		74

Course ID	Phase	Course Level	MOS	School
				RTS-M-02 CAMP DODGE IA
				RTS-M-04 FT BRAGG NC
				RTS-M-12 FT STEWART GA
	(blank)	MOS	45k1	RTS-M-09 GOWEN FIELD ID
091-45K10 Total				
091-45K10 (T)	(blank)	Transition	45K	ORDNANCE SCH, APG
				RTS-M-06 CAMP ROBERT CA
091-45K10 (T) Total				
091-45K30	2	BNCOC	45k3	NCO ACADEMY - APG
				RTS-M, FT HOOD, TX
				RTS-M-02 CAMP DODGE IA
				RTS-M-04 FT BRAGG NC
	3	BNCOC	45k3	NCO ACADEMY - APG
				RTS-M, FT HOOD, TX
				RTS-M-02 CAMP DODGE IA
				RTS-M-04 FT BRAGG NC
091-45K30 Total				
091-45K40	2	ANCOC	45K4	NCO ACADEMY - APG
				RTS-M, FT MCCOY, WI
				RTS-M-10 BLANDING FL
091-45K40 Total				
091-45T10	1	MOS	45t1	RTS-M-12 FT STEWART GA
	2	MOS	45t1	RTS-M-12 FT STEWART GA
091-45T10 Total				
091-52C10	1	MOS	52c1	Jefferson City RTSM
				ORDNANCE SCH, APG
				RTS-M-01 SALINA KS
				RTS-M-02 CAMP DODGE IA
				RTS-M-12 FT STEWART GA
	2	MOS	52c1	Jefferson City RTSM
				ORDNANCE SCH, APG
				RTS-M-01 SALINA KS
091-52C10 Total				
091-52C30	2	BNCOC	52c3	Jefferson City RTSM
				NCO ACADEMY - APG
				RTS-M-01 SALINA KS
091-52C30 Total				
091-52D10	1	MOS	52d1	ORDNANCE SCH, APG
				RTS-M, FT HOOD, TX
				RTS-M-02 CAMP DODGE IA
				RTS-M-03 FT DIX NJ

Base Case		Nearest School		Reassign Courses Multifunctional		Reassign Courses Specialized		Consolidate Schools Specialized	
AC	RC	AC	RC	AC	RC	AC	RC	AC	RC
					3		3		3
	3		3						
	1		1		1		1		1
	4		4		4		4		4
							8		8
	8		8		8				
			8		8		8		8
	8								
	16		16		16		16		16
					5		6		6
	3		3		5				
	12		12		5		9		9
					5		12		12
	3								
	5		12						
	4				7				
			16		5		20		20
	4		4		5				
	16				10				
	47		47		47		47		47
			11		5		11		11
	11				6				
			20		6		20		20
	20				14				
	31		31		31		31		31
	4		4		4		4		4
	3				9		9		9
	6		9						
			25		13				
	25				12		25		25
	38		38		38		38		38
			7		8		8		8
	4		5		5				
	47		45		26				
	6				18		49		49
							12		12
	6		25		10				
	30				27				
	6		6						
	6		17		11		36		36
			26		5				

Course ID	Phase	Course Level	MOS	School
091-44E10	1	MOS	44E10	ORDNANCE SCH, APG
				RTS-M-01 SALINA KS
	2	MOS	44E10	RTS-M-01 SALINA KS
091-44E10 Total				
091-44E30	2	BNCOC	44E30	NCO ACADEMY - APG
				RTS-M-01 SALINA KS
	3	BNCOC	44E30	NCO ACADEMY - APG
				RTS-M-01 SALINA KS
091-44E30 Total				
091-45B10	1	MOS	45b1	ORDNANCE SCH, APG
				RTS-M, FT HOOD, TX
				RTS-M-01 SALINA KS
	2	MOS	45b1	ORDNANCE SCH, APG
				RTS-M, FT HOOD, TX
				RTS-M-01 SALINA KS
				RTS-M-02 CAMP DODGE IA
	(blank)	MOS	45b1	ORDNANCE SCH, APG
				RTS-M-09 GOWEN FIELD ID
				RTS-M-10 BLANDING FL
091-45B10 Total				
091-45D10	1	MOS	45d1	FLD ARTILLERY SCH, SILL
				RTS-M-07 CAMP RIPLEY MN
	2	MOS	45d1	FLD ARTILLERY SCH, SILL
				RTS-M-07 CAMP RIPLEY MN
091-45D10 Total				
091-45E10	1	MOS	45E10	RTS-M-06 CAMP ROBERT CA
	2	MOS	45E10	RTS-M-06 CAMP ROBERT CA
				RTS-M-07 CAMP RIPLEY MN
	(blank)	MOS	45E10	ARMOR SCH, FT KNOX
				RTS-M-09 GOWEN FIELD ID
091-45E10 Total				
091-45K10	1	MOS	45k1	ORDNANCE SCH, APG
				RTS-M, FT HOOD, TX
				RTS-M-02 CAMP DODGE IA
				RTS-M-12 FT STEWART GA
	2	MOS	45k1	ORDNANCE SCH, APG
				RTS-M, FT INDNTWN GP PA
				RTS-M-02 CAMP DODGE IA
				RTS-M-10 BLANDING FL
				RTS-M-12 FT STEWART GA
	3	MOS	45k1	ORDNANCE SCH, APG

Courses Taught at Each School for the Various Options

Base Case		Nearest School		Reassign Courses Multifunctional		Reassign Courses Specialized		Consolidate Schools Specialized	
AC	RC	AC	RC	AC	RC	AC	RC	AC	RC
	14				5				
	8				5		26		26
	4				11				
	7		7		7		7		7
	138		138		138		138		138
							14		14
	14		14		14				
	14		14		14		14		14
					6				
	3		3						
	10		3		5		16		16
	3		10		5				
			2						
	7				7				
	7		10		5		17		17
	3		5		5				
	33		33		33		33		33
					6				
	15		15		5				
	1		1		5		16		16
	16		16		16		16		16
	6		6		6		6		6
	2		2		2		2		2
	8		8		8		8		8
	9		12		28				
			43		17				
	25				15				25
	21				6		66		
	11		11						41
	10		12		19				
			43		6		55		13
	45				30				42
	121		121		121		121		121
	6		16		5		16		16
					6				
	10				5				
	16		16		16		16		16
			21		5				
	4		15		6		14		
	44		14		13		34		39
	11		7		5				

Course ID	Phase	Course Level	MOS	School
				RTS-M-04 FT BRAGG NC
				RTS-M-06 CAMP ROBERT CA
				RTS-M-08 FT CUSTER MI
				RTS-M-12 FT STEWART GA
	2	MOS	52d1	Ft. Devens RTSM
				Jefferson City RTSM
				ORDNANCE SCH, APG
				RTS-M, FT HOOD, TX
				RTS-M, FT MCCOY, WI
				RTS-M-02 CAMP DODGE IA
				RTS-M-03 FT DIX NJ
				RTS-M-04 FT BRAGG NC
				RTS-M-05 CAMP SHELBY MS
				RTS-M-06 CAMP ROBERT CA
				RTS-M-08 FT CUSTER MI
				RTS-M-10 BLANDING FL
				RTS-M-12 FT STEWART GA
	(blank)	MOS	52d1	ORDNANCE SCH, APG
				RTS-M-10 BLANDING FL
091-52D10 Total				
091-52D30	2	BNCOC	52d3	Jefferson City RTSM
				NCO ACADEMY - APG
				RTS-M, FT HOOD, TX
				RTS-M, FT MCCOY, WI
				RTS-M-02 CAMP DODGE IA
				RTS-M-03 FT DIX NJ
				RTS-M-04 FT BRAGG NC
				RTS-M-05 CAMP SHELBY MS
091-52D30 Total				
091-52X40	2	ANCOC	52X4	ORDNANCE SCH, APG
				RTS-M, FT INDNTWN GP PA
091-52X40 Total				
091-62B40	2	ANCOC	62B4	RTS-M, FT HOOD, TX
				RTS-M, FT MCCOY, WI
				RTS-M-09 GOWEN FIELD ID
091-62B40 Total				
091-63B/S/W10H8	(blank)	ASI	63BSWasi	Jefferson City RTSM
				ORDNANCE SCH, APG
				RTS-M, FT HOOD, TX
				RTS-M, FT INDNTWN GP PA
				RTS-M, FT MCCOY, WI

Courses Taught at Each School for the Various Options

Base Case		Nearest School		Reassign Courses Multifunctional		Reassign Courses Specialized		Consolidate Schools Specialized	
AC	RC	AC	RC	AC	RC	AC	RC	AC	RC
	9				17		56		65
	24		22		22				
	4		17		8				
	8		8		28				
	6		6		6				
	6		24		5				
			2		5				
	5		8		7				
	4		2		8		45		
	25		9		7				
	9		7		5				14
	12		7		7				39
	13		15		9		50		42
	18		17		17		21		21
	10		17		16				
	7		1						
	1		1		24				
					5		11		11
	11		11		6				
	231		231		231		231		231
	7		7		7		28		
	10		13		13		30		
	9		9		14				
	9		5						
	6		5		5				
	6		6		6				58
	11		13		13				
	58		58		58		58		58
			7		7		12		
	12		5		5				12
	12		12		12		12		12
	4		4		5				
	5		5		5				
	7		7		6		16		16
	16		16		16		16		16
	7		12		24				
			11		11				
	11		10		5		10		18
	7		8		8				
	16		11		11				

Course ID	Phase	Course Level	MOS	School
				RTS-M-02 CAMP DODGE IA
				RTS-M-03 FT DIX NJ
				RTS-M-04 FT BRAGG NC
				RTS-M-06 CAMP ROBERT CA
				RTS-M-07 CAMP RIPLEY MN
				RTS-M-10 BLANDING FL
				Waiwa RTSM
091-63B/S/W10H8 Tot				
091-63B10	1	MOS	63b1	Jefferson City RTSM
				RTS-M, FT HOOD, TX
				RTS-M-01 SALINA KS
				RTS-M-02 CAMP DODGE IA
				RTS-M-03 FT DIX NJ
				RTS-M-04 FT BRAGG NC
				RTS-M-06 CAMP ROBERT CA
				RTS-M-08 FT CUSTER MI
				RTS-M-09 GOWEN FIELD ID
				RTS-M-10 BLANDING FL
				RTS-M-12 FT STEWART GA
				USATC, FT. JACKSON/108TH
	2	MOS	63b1	Jefferson City RTSM
				RTS-M, FT HOOD, TX
				RTS-M, FT INDNTWN GP PA
				RTS-M, FT MCCOY, WI
				RTS-M-01 SALINA KS
				RTS-M-02 CAMP DODGE IA
				RTS-M-03 FT DIX NJ
				RTS-M-04 FT BRAGG NC
				RTS-M-05 CAMP SHELBY MS
				RTS-M-06 CAMP ROBERT CA
				RTS-M-07 CAMP RIPLEY MN
				RTS-M-08 FT CUSTER MI
				RTS-M-09 GOWEN FIELD ID
				RTS-M-10 BLANDING FL
				RTS-M-12 FT STEWART GA
				USATC, FT. JACKSON/108TH
				Waiwa RTSM
091-63B10 Total				
091-63B30	2	BNCOC	63b3	Jefferson City RTSM
				NCO ACADEMY - APG
				RTS-M, FT HOOD, TX

Courses Taught at Each School for the Various Options

Base Case		Nearest School		Reassign Courses Multifunctional		Reassign Courses Specialized		Consolidate Schools Specialized	
AC	RC	AC	RC	AC	RC	AC	RC	AC	RC
	30		22		20		60		
	12		5		5		47		47
	15		15		12				
	13		13		15		24		15
	10		9		9				52
	11		16		12				
	9		9		9				9
	141		141		141		141		141
	4		21		11				
	2		5		5		5		5
	19		9		9				22
	17		14		14		33		
	11		20		20				37
	20		22		22				
	57		41		41		41		41
	32		32		42		59		53
	32		36		36		36		36
	11		10		10				37
	26		11		11				
			10		10		57		
	28		51		36				36
	35		35		26		56		26
	18		24		33				
	35		33		33		68		
	27		27		33		91		33
	15		20		23		64		23
	33		32		23		56		56
	23		18		18				
	54		32		32				32
	64		45		45				
	43		16		31				45
	39		52		56		56		75
	32		36		36				81
	11		9		9		25		29
	29		47		32				
			9		20		70		50
	6		6		6		6		6
	723		723		723		723		723
	19		32		25				
			13		13				
	14		17		9		17		17

Course ID	Phase	Course Level	MOS	School
				RTS-M, FT INDNTWN GP PA
				RTS-M, FT MCCOY, WI
				RTS-M-01 SALINA KS
				RTS-M-02 CAMP DODGE IA
				RTS-M-03 FT DIX NJ
				RTS-M-04 FT BRAGG NC
				RTS-M-05 CAMP SHELBY MS
				RTS-M-06 CAMP ROBERT CA
				RTS-M-07 CAMP RIPLEY MN
				RTS-M-08 FT CUSTER MI
				RTS-M-09 GOWEN FIELD ID
				RTS-M-10 BLANDING FL
091-63B30 Total				
091-63B40	2	ANCOC	63B4	Jefferson City RTSM
				NCO ACADEMY - APG
				RTS-M, FT HOOD, TX
				RTS-M, FT INDNTWN GP PA
				RTS-M, FT MCCOY, WI
				RTS-M-01 SALINA KS
				RTS-M-02 CAMP DODGE IA
				RTS-M-03 FT DIX NJ
				RTS-M-05 CAMP SHELBY MS
				RTS-M-06 CAMP ROBERT CA
				RTS-M-07 CAMP RIPLEY MN
				RTS-M-09 GOWEN FIELD ID
				RTS-M-10 BLANDING FL
091-63B40 Total				
091-63D/E/H/N/T/Y10H8	(blank)	ASI	63DEHNTY asi	RTS-M, FT HOOD, TX
				RTS-M, FT INDNTWN GP PA
				RTS-M, FT MCCOY, WI
				RTS-M-02 CAMP DODGE IA
				RTS-M-03 FT DIX NJ
				RTS-M-07 CAMP RIPLEY MN
				RTS-M-08 FT CUSTER MI
091-63D/E/H/N/T/Y10H8 Total				
091-63D10	1	MOS	63d1	ORDNANCE SCH, APG
				RTS-M, FT MCCOY, WI
				RTS-M-01 SALINA KS
				RTS-M-12 FT STEWART GA
	2	MOS	63d1	ORDNANCE SCH, APG

Courses Taught at Each School for the Various Options

Base Case		Nearest School		Reassign Courses Multifunctional		Reassign Courses Specialized		Consolidate Schools Specialized	
AC	RC	AC	RC	AC	RC	AC	RC	AC	RC
	9		27		17				
	49		16		16				
	5		14		21				
	27		10		11		43		52
	32		24		24		54		54
	25		16		10		10		10
	40		46		37		44		29
	27		18		18		18		18
	23		29		27				
	10		14		39		72		66
	14		18		18		27		18
	4		4		13		13		34
	298		298		298		298		298
	21		28		19				
			18		13				
	8		8		8				15
	6		12		12		39		39
	38		14		9				
	10		19		17				
	12		12		11				
	17		14		14				
	51		30		22		50		17
	14		14		14				14
	15		16		36		101		71
	8		15		15				21
	6		6		16		16		29
	206		206		206		206		206
	10		5		5		10		10
	10		10		21				
	5		11		12				
	5		5		9				
	18		18		7		56		56
	11		10		5				
	7		7		7				
	66		66		66		66		66
			5		5		11		11
	2		2		30				
	34						38		38
	13		42		14				
			43		5				

Course ID	Phase	Course Level	MOS	School
				RTS-M, FT HOOD, TX
				RTS-M, FT INDNTWN GP PA
				RTS-M-01 SALINA KS
				RTS-M-05 CAMP SHELBY MS
				RTS-M-07 CAMP RIPLEY MN
				RTS-M-12 FT STEWART GA
	(blank)	MOS	63d1	ORDNANCE SCH, APG
				RTS-M-12 FT STEWART GA
091-63D10 Total				
091-63D30 TRK II	2	BNCOC	63d3	NCO ACADEMY - APG
				RTS-M-01 SALINA KS
091-63D30 TRK II Total				
091-63D40	2	ANCOC	63D4	NCO ACADEMY - APG
				RTS-M-01 SALINA KS
091-63D40 Total				
091-63E10	1	MOS	63E10	ARMOR SCH, FT KNOX
				RTS-M-01 SALINA KS
				RTS-M-06 CAMP ROBERT CA
				RTS-M-08 FT CUSTER MI
				RTS-M-12 FT STEWART GA
	2	MOS	63E10	ARMOR SCH, FT KNOX
				RTS-M, FT INDNTWN GP PA
				RTS-M-01 SALINA KS
				RTS-M-02 CAMP DODGE IA
				RTS-M-06 CAMP ROBERT CA
				RTS-M-07 CAMP RIPLEY MN
				RTS-M-08 FT CUSTER MI
	(blank)	MOS	63E10	ARMOR SCH, FT KNOX
				RTS-M-09 GOWEN FIELD ID
091-63E10 Total				
091-63E30	2	BNCOC	63E30	NCO ACADEMY - FT KNOX
				RTS-M, FT INDNTWN GP PA
				RTS-M-09 GOWEN FIELD ID
091-63E30 Total				
091-63E40	2	ANCOC	63E40	NCO ACADEMY - FT KNOX
				RTS-M-01 SALINA KS
				RTS-M-09 GOWEN FIELD ID
091-63E40 Total				
091-63G10	1	MOS	63g1	RTS-M-01 SALINA KS
	2	MOS	63g1	ORDNANCE SCH, APG
				RTS-M-01 SALINA KS

Courses Taught at Each School for the Various Options   57

Base Case		Nearest School		Reassign Courses Multifunctional		Reassign Courses Specialized		Consolidate Schools Specialized	
AC	RC	AC	RC	AC	RC	AC	RC	AC	RC
	24		24		54				
	6		15		25				
	21						33		33
	9		25		14				
	32						49		49
	15				9		25		25
			20		8		20		20
	20				12				
	176		176		176		176		176
					5		12		12
	12		12		7				
	12		12		12		12		12
					8		8		8
	8		8						
	8		8		8		8		8
					5				
	4		9		21				
	21		27				20		20
	11				10				22
	6		6		6		22		
			48		5				
	7		8		7		42		42
	3		9		23				
	9				9				
	23						23		23
	16				16				
	7				5				
			39		23				
	39				16		39		39
	146		146		146		146		146
			12				12		12
	6				5				
	6				7				
	12		12		12		12		12
					13				
	10		16		5				
	13		7		5		23		23
	23		23		23		23		23
	10		10		10		10		10
					5		11		11
	11		11		6				

58 Consolidating Active and Reserve Component Training Infrastructure

Course ID	Phase	Course Level	MOS	School
091-63G10 Total				
091-63H10	1	MOS	63h1	Jefferson City RTSM
				RTS-M, FT HOOD, TX
				RTS-M-01 SALINA KS
				RTS-M-02 CAMP DODGE IA
				RTS-M-03 FT DIX NJ
				RTS-M-04 FT BRAGG NC
				RTS-M-05 CAMP SHELBY MS
				RTS-M-06 CAMP ROBERT CA
				RTS-M-08 FT CUSTER MI
				RTS-M-12 FT STEWART GA
	2	MOS	63h1	Jefferson City RTSM
				RTS-M, FT HOOD, TX
				RTS-M, FT INDNTWN GP PA
				RTS-M, FT MCCOY, WI
				RTS-M-01 SALINA KS
				RTS-M-02 CAMP DODGE IA
				RTS-M-03 FT DIX NJ
				RTS-M-04 FT BRAGG NC
				RTS-M-05 CAMP SHELBY MS
				RTS-M-06 CAMP ROBERT CA
				RTS-M-07 CAMP RIPLEY MN
				RTS-M-08 FT CUSTER MI
				RTS-M-12 FT STEWART GA
	(blank)	MOS	63h1	ORDNANCE SCH, APG
				RTS-M-09 GOWEN FIELD ID
091-63H10 Total				
091-63H10 (GS) BFV	1	MOS	63h1	RTS-M-01 SALINA KS
	2	MOS	63h1	ORDNANCE SCH, APG
				RTS-M-01 SALINA KS
091-63H10 (GS) BFV Total				
091-63H10 (GS) M1	(blank)	MOS	63h1	ORDNANCE SCH, APG
				RTS-M-02 CAMP DODGE IA
091-63H10 (GS) M1 Total				
091-63H10 (T)	(blank)	Transition	63H	ORDNANCE SCH, APG
				RTS-M-03 FT DIX NJ
				RTS-M-06 CAMP ROBERT CA
091-63H10 (T) Total				
091-63H10L8	(blank)	ASI	63H	ORDNANCE SCH, APG
				RTS-M-02 CAMP DODGE IA
091-63H10L8 Total				

Courses Taught at Each School for the Various Options

Base Case		Nearest School		Reassign Courses Multifunctional		Reassign Courses Specialized		Consolidate Schools Specialized	
AC	RC	AC	RC	AC	RC	AC	RC	AC	RC
	21		21		21		21		21
	5		9		42				
	14		8		5		5		5
	3		6		5		15		8
	49		12		10				
	7		7		7		22		22
	12		15		28				
	10		34		35		87		87
	25		30		36		36		36
	41		33		36		39		46
	38		50						
	4		12		6		6		
	19		21		31		32		52
	25		25		33		50		38
	3		16		16				
	33		18		15		44		
	52		5		7				49
	4		12		5				
	4		11		10				10
	28		36		14		14		23
	20		27		27				
	22		21		28		63		
	30		23		23				51
	33		50		62		68		54
			23		11		23		23
	23				12				
	504		504		504		504		504
	3		3		3		3		3
					5		5		5
	5		5						
	8		8		8		8		8
			8		8		8		8
	8								
	8		8		8		8		8
			11						
	10				11		11		11
	49		48		48		48		48
	59		59		59		59		59
					6		6		6
	6		6						
	6		6		6		6		6

Course ID	Phase	Course Level	MOS	School
091-63H30	2	BNCOC	63h3	Ft. Devens RTSM
				RTS-M, FT HOOD, TX
				RTS-M, FT INDNTWN GP PA
				RTS-M, FT MCCOY, WI
				RTS-M-04 FT BRAGG NC
				RTS-M-06 CAMP ROBERT CA
				RTS-M-07 CAMP RIPLEY MN
				RTS-M-09 GOWEN FIELD ID
				RTS-M-12 FT STEWART GA
091-63H30 Total				
091-63H30-IIA	2	BNCOC	63h3	Jefferson City RTSM
				NCO ACADEMY - APG
				RTS-M-02 CAMP DODGE IA
				RTS-M-05 CAMP SHELBY MS
091-63H30-IIA Total				
091-63H30-IIB	2	BNCOC	63h3	Jefferson City RTSM
				NCO ACADEMY - APG
				RTS-M-02 CAMP DODGE IA
091-63H30-IIB Total				
091-63H40	2	ANCOC	63H4	NCO ACADEMY - APG
				RTS-M, FT HOOD, TX
				RTS-M, FT INDNTWN GP PA
				RTS-M-01 SALINA KS
				RTS-M-02 CAMP DODGE IA
				RTS-M-05 CAMP SHELBY MS
				RTS-M-06 CAMP ROBERT CA
				RTS-M-07 CAMP RIPLEY MN
				RTS-M-09 GOWEN FIELD ID
				RTS-M-12 FT STEWART GA
091-63H40 Total				
091-63J10	1	MOS	63j1	ORDNANCE SCH, APG
				RTS-M, FT INDNTWN GP PA
				RTS-M-02 CAMP DODGE IA
				RTS-M-12 FT STEWART GA
	2	MOS	63j1	ORDNANCE SCH, APG
				RTS-M, FT INDNTWN GP PA
091-63J10 Total				
091-63S10	1	MOS	63s1	Jefferson City RTSM
				RTS-M-01 SALINA KS
				RTS-M-02 CAMP DODGE IA
				RTS-M-03 FT DIX NJ

Courses Taught at Each School for the Various Options   61

Base Case		Nearest School		Reassign Courses Multifunctional		Reassign Courses Specialized		Consolidate Schools Specialized	
AC	RC	AC	RC	AC	RC	AC	RC	AC	RC
	4		4		5				
	23		23		22		46		16
	5		5		5		33		63
	10		10		10				
	4		4		5				
	4		4		5				
	11		11		10				
	4		4		5				
	14		14		12				
	79		79		79		79		79
	5				5				
			12		10				
	27				14		38		
	6		26		9				38
	38		38		38		38		38
	12		12		10				40
					13		40		
	28		28		17				
	40		40		40		40		40
			5		5				
	10		10		8				
	5		10		10		24		35
	13		13		11		5		49
	23		9		11				
	14		7		7				
	9		5		5				
	6		15		17				
	1		5		5		64		9
	12		14		14				
	93		93		93		93		93
			50		5				
	9		9		17		27		41
	50				28				31
	13		13		22		45		
			26		14		26		
	26				12				26
	98		98		98		98		98
	6		6		5				53
	19		9		9		30		
	2		5		5				
	2		3		12				

Course ID	Phase	Course Level	MOS	School
				RTS-M-04 FT BRAGG NC
				RTS-M-05 CAMP SHELBY MS
				RTS-M-06 CAMP ROBERT CA
				RTS-M-08 FT CUSTER MI
				RTS-M-09 GOWEN FIELD ID
				RTS-M-12 FT STEWART GA
				USATC, FT. JACKSON/108TH
	2	MOS	63s1	Jefferson City RTSM
				RTS-M, FT HOOD, TX
				RTS-M, FT INDNTWN GP PA
				RTS-M, FT MCCOY, WI
				RTS-M-01 SALINA KS
				RTS-M-04 FT BRAGG NC
				RTS-M-05 CAMP SHELBY MS
				RTS-M-06 CAMP ROBERT CA
				RTS-M-07 CAMP RIPLEY MN
				RTS-M-08 FT CUSTER MI
				RTS-M-09 GOWEN FIELD ID
				RTS-M-12 FT STEWART GA
				USATC, FT. JACKSON/108TH
091-63S10 Total				
091-63T10	1	MOS	63t1	ARMOR SCH, FT KNOX
				RTS-M-01 SALINA KS
				RTS-M-03 FT DIX NJ
				RTS-M-04 FT BRAGG NC
				RTS-M-06 CAMP ROBERT CA
				RTS-M-08 FT CUSTER MI
				RTS-M-09 GOWEN FIELD ID
				RTS-M-12 FT STEWART GA
	2	MOS	63t1	ARMOR SCH, FT KNOX
				RTS-M, FT HOOD, TX
				RTS-M, FT INDNTWN GP PA
				RTS-M-01 SALINA KS
				RTS-M-03 FT DIX NJ
				RTS-M-04 FT BRAGG NC
				RTS-M-05 CAMP SHELBY MS
				RTS-M-06 CAMP ROBERT CA
				RTS-M-07 CAMP RIPLEY MN
				RTS-M-08 FT CUSTER MI
				RTS-M-09 GOWEN FIELD ID
				RTS-M-12 FT STEWART GA

Courses Taught at Each School for the Various Options

Base Case		Nearest School		Reassign Courses Multifunctional		Reassign Courses Specialized		Consolidate Schools Specialized	
AC	RC	AC	RC	AC	RC	AC	RC	AC	RC
	13		22		14				45
	9		3		5				
	15		15		15		25		25
	16		25		27				
	25		12		12				
	16		20		14				
			3		5		68		
	12		11		11				
	5		3		5				6
	5		9		15		23		30
	11		3		6				
	19		22		17		37		
	16		14		8				
	16		10		10				
	10		11		11		11		11
	9		14		14				55
	15		20		28		44		
	22		11		11		11		12
	20		29		14		34		46
			3		10				
	283		283		283		283		283
					5				
	12		35		9				24
	14		14		11		49		58
	27				37				
	16		20		7		8		
	8		14		10		32		
	17		24		19		18		25
	13				9				
			5		5				
	5		5		5		7		12
	38		38				49		
	26		23		23		28		19
	33		33		71		22		72
	8		9		9		20		
	7		10		10				
	10		6		6		18		6
	18		20		20		20		23
	7		6		6		11		
	17		16		16				16
	23		21		21		17		44

Course ID	Phase	Course Level	MOS	School
091-63T10 Total				
091-63T30	2	BNCOC	63t3	NCO ACADEMY - FT KNOX
				RTS-M, FT HOOD, TX
				RTS-M-04 FT BRAGG NC
				RTS-M-07 CAMP RIPLEY MN
				RTS-M-09 GOWEN FIELD ID
091-63T30 Total				
091-63T40	2	ANCOC	63T4	NCO ACADEMY - FT KNOX
				RTS-M-01 SALINA KS
091-63T40 Total				
091-63W10	1	MOS	63w1	Jefferson City RTSM
				ORDNANCE SCH, APG
				RTS-M, FT INDNTWN GP PA
				RTS-M, FT MCCOY, WI
				RTS-M-01 SALINA KS
				RTS-M-02 CAMP DODGE IA
				RTS-M-03 FT DIX NJ
				RTS-M-04 FT BRAGG NC
				RTS-M-06 CAMP ROBERT CA
				RTS-M-12 FT STEWART GA
	2	MOS	63w1	Jefferson City RTSM
				ORDNANCE SCH, APG
				RTS-M, FT HOOD, TX
				RTS-M, FT INDNTWN GP PA
				RTS-M, FT MCCOY, WI
				RTS-M-01 SALINA KS
				RTS-M-02 CAMP DODGE IA
				RTS-M-03 FT DIX NJ
				RTS-M-04 FT BRAGG NC
				RTS-M-06 CAMP ROBERT CA
				RTS-M-07 CAMP RIPLEY MN
				RTS-M-10 BLANDING FL
				RTS-M-12 FT STEWART GA
	(blank)	MOS	63w1	ORDNANCE SCH, APG
				RTS-M-09 GOWEN FIELD ID
091-63W10 Total				
091-63W10 (GS) HEMTT	1	MOS	63w1	ORDNANCE SCH, APG
				RTS-M-01 SALINA KS
091-63W10 (GS) HEMTT Total				

Base Case		Nearest School		Reassign Courses Multifunctional		Reassign Courses Specialized		Consolidate Schools Specialized	
AC	RC	AC	RC	AC	RC	AC	RC	AC	RC
	299		299		299		299		299
					7				
	5								
	7		10		5				
	3		13		5		23		23
	8				6				
	23		23		23		23		23
			11		5		11		11
	11				6				
	11		11		11		11		11
	25		38		23				68
			36		5				37
	5		11		8				
	7		20		29		44		
	24		31		14		38		
	85		30		6				
	12		7		7		20		
	14		14		19		81		
	15		28		35		32		35
	28				69				75
	17		38		20		26		20
			12		5				
	13		25		9		9		13
	21		24		31		41		39
	17		10		10				
	27		19		12		13		9
	78		49		28				
	16		12		5				
	9		9		14		14		22
	15		28		32		32		32
	41		51		45		82		82
	21		40		18		41		41
	42				88		59		59
									21
	21		21		21		21		
	553		553		553		553		553
			12		5		12		12
	12				7				
	12		12		12		12		12

Course ID	Phase	Course Level	MOS	School
091-63W10 (GS) HMMWV	1	MOS	63w1	ORDNANCE SCH, APG
				RTS-M-01 SALINA KS
	2	MOS	63w1	ORDNANCE SCH, APG
				RTS-M-01 SALINA KS
091-63W10 (GS) HMMWV Total				
091-63W10 (GS) M939	1	MOS	63w1	ORDNANCE SCH, APG
				RTS-M-01 SALINA KS
	(blank)	MOS	63w1	RTS-M-02 CAMP DODGE IA
091-63W10 (GS) M939 Total				
091-63Y10	1	MOS	63y1	ORDNANCE SCH, APG
				RTS-M-01 SALINA KS
				RTS-M-03 FT DIX NJ
				RTS-M-08 FT CUSTER MI
				RTS-M-12 FT STEWART GA
	2	MOS	63y1	ORDNANCE SCH, APG
				RTS-M, FT HOOD, TX
				RTS-M-01 SALINA KS
				RTS-M-03 FT DIX NJ
				RTS-M-08 FT CUSTER MI
				RTS-M-09 GOWEN FIELD ID
				RTS-M-12 FT STEWART GA
091-63Y10 Total				
091-ASIL8 (45K)	1	ASI	45Kasi	ORDNANCE SCH, APG
				RTS-M-03 FT DIX NJ
091-ASIL8 (45K) Total				
093-27E10	1	MOS	27e	RTS-M-02 CAMP DODGE IA
	2	MOS	27e	RTS-M-02 CAMP DODGE IA
	3	MOS	27e	RTS-M-02 CAMP DODGE IA
	4	MOS	27e	RTS-M-02 CAMP DODGE IA
093-27E10 Total				
093-27E30	2	BNCOC	27e	RTS-M-02 CAMP DODGE IA
093-27E30 Total				
101-92A10	1	MOS	92a1	Jefferson City RTSM
				QUARTERMASTER, FT LEE
				RTS-M-03 FT DIX NJ
				RTS-M-04 FT BRAGG NC
				RTS-M-06 CAMP ROBERT CA
				RTS-M-12 FT STEWART GA

Courses Taught at Each School for the Various Options

Base Case		Nearest School		Reassign Courses Multifunctional		Reassign Courses Specialized		Consolidate Schools Specialized	
AC	RC	AC	RC	AC	RC	AC	RC	AC	RC
					3		3		3
	3		3						
					3		3		3
	3		3						
	6		6		6		6		6
					8		8		8
	8		8						
	8		8		8		8		8
	16		16		16		16		16
					13				
	12		9		5				
	21		19		10		49		49
	12		12		16				
	4		9		5				
			9		9				
	7		5		5				
	6		5		5				
	15		8		8		50		50
	8		7		8				
	3		5		5				
	11		11		10				
	99		99		99		99		99
							6		6
	6		6		6				
	6		6		6		6		6
	20		20		20		20		20
	18		18		18		18		18
	9		9		9		9		9
	9		9		9		9		9
	56		56		56		56		56
	6		6		6		6		6
	6		6		6		6		6
	31		40		54		64		65
					7				
	46		113		47		52		54
	26		64		34		39		107
	42		53		46		44		44
	123				82		71		

Course ID	Phase	Course Level	MOS	School
				Waiwa RTSM
	2	MOS	92a1	Jefferson City RTSM
				QUARTERMASTER, FT LEE
				RTS-M-03 FT DIX NJ
				RTS-M-04 FT BRAGG NC
				RTS-M-05 CAMP SHELBY MS
				RTS-M-06 CAMP ROBERT CA
				RTS-M-07 CAMP RIPLEY MN
				RTS-M-10 BLANDING FL
				RTS-M-12 FT STEWART GA
				Waiwa RTSM
101-92A10 Total				
101-92A30	2	BNCOC	92a3	Jefferson City RTSM
				NCO ACADEMY - FT LEE
				RTS-M-07 CAMP RIPLEY MN
				RTS-M-12 FT STEWART GA
				Waiwa RTSM
101-92A30 Total				
101-92A40	2	ANCOC	92a4	Jefferson City RTSM
				NCO ACADEMY - FT LEE
				RTS-M-07 CAMP RIPLEY MN
				RTS-M-12 FT STEWART GA
101-92A40 Total				
113-45G10	(blank)	MOS	45g1	ORDNANCE SCH, APG
113-45G10 Total				
551-92A10	(blank)	MOS	92a1	Jefferson City RTSM
				QUARTERMASTER, FT LEE
				RTS-M-03 FT DIX NJ
				RTS-M-04 FT BRAGG NC
				RTS-M-05 CAMP SHELBY MS
				RTS-M-06 CAMP ROBERT CA
				RTS-M-07 CAMP RIPLEY MN
				RTS-M-10 BLANDING FL
				RTS-M-12 FT STEWART GA
				Waiwa RTSM
551-92A10 Total				
551-92A30	(blank)	BNCOC	92a3	Jefferson City RTSM
				NCO ACADEMY - FT LEE
				RTS-M-03 FT DIX NJ
				RTS-M-04 FT BRAGG NC
				RTS-M-05 CAMP SHELBY MS

Courses Taught at Each School for the Various Options

Base Case		Nearest School		Reassign Courses Multifunctional		Reassign Courses Specialized		Consolidate Schools Specialized	
AC	RC	AC	RC	AC	RC	AC	RC	AC	RC
	13		11		11		11		11
	24		40		29		47		36
			10		9				
	79		70		70		110		79
	35		43		43				43
	100		40		40				
	39		50		50		49		49
	39		39		51		51		51
	49		49		49		103		93
	77		104		104		85		94
	28		25		25		25		25
	751		751		751		751		751
	21				9		47		47
			9		5				
	15		27		12				
	6		6		16				
	5		5		5				
	47		47		47		47		47
	15		20		5				
			2		5				
	7				9				
	2		2		5		24		24
	24		24		24		24		24
40		40		40		40		40	
40		40		40		40		40	
			14		5		14		14
231			28				100		100
			5		33				
			48		48				
			57		57		57		57
			14		14		36		36
					28				
					24		24		24
			43						
			22		22				
231		231		231		231		231	
			40		8		70		51
252							84		99
			63		16				
					47		15		
					54				

Course ID	Phase	Course Level	MOS	School
				RTS-M-06 CAMP ROBERT CA
				RTS-M-07 CAMP RIPLEY MN
				RTS-M-10 BLANDING FL
				RTS-M-12 FT STEWART GA
				Waiwa RTSM
551-92A30 Total				
551-92A40	(blank)	ANCOC	92a4	Jefferson City RTSM
				NCO ACADEMY - FT LEE
				RTS-M-07 CAMP RIPLEY MN
				RTS-M-12 FT STEWART GA
				Waiwa RTSM
551-92A40 Total				
6-63-C42	(blank)	ANCOC	63-all	Ft. Devens RTSM
				Jefferson City RTSM
				NCO ACADEMY - APG
				RTS-M, FT HOOD, TX
				RTS-M, FT INDNTWN GP PA
				RTS-M, FT MCCOY, WI
				RTS-M-01 SALINA KS
				RTS-M-02 CAMP DODGE IA
				RTS-M-04 FT BRAGG NC
				RTS-M-05 CAMP SHELBY MS
				RTS-M-06 CAMP ROBERT CA
				RTS-M-07 CAMP RIPLEY MN
				RTS-M-08 FT CUSTER MI
				RTS-M-09 GOWEN FIELD ID
				RTS-M-10 BLANDING FL
				RTS-M-12 FT STEWART GA
				Waiwa RTSM
6-63-C42 Total				
610-63B10	(blank)	MOS	63b1	Jefferson City RTSM
				RTS-M, FT HOOD, TX
				RTS-M, FT INDNTWN GP PA
				RTS-M, FT MCCOY, WI
				RTS-M-02 CAMP DODGE IA
				RTS-M-03 FT DIX NJ
				RTS-M-04 FT BRAGG NC
				RTS-M-05 CAMP SHELBY MS
				RTS-M-08 FT CUSTER MI
				RTS-M-09 GOWEN FIELD ID
				RTS-M-10 BLANDING FL

Courses Taught at Each School for the Various Options

Base Case		Nearest School		Reassign Courses Multifunctional		Reassign Courses Specialized		Consolidate Schools Specialized	
AC	RC	AC	RC	AC	RC	AC	RC	AC	RC
		41		34		34		34	
		40		46					
		56		30		37		56	
				5					
		12		12		12		12	
252		252		252		252		252	
				26					
135				74		80		81	
		20		21		48		47	
				7					
		115		7		7		7	
135		135		135		135		135	
		5		5		13			
		26		26		44		42	
520		60		60		67		104	
		84		84		84		84	
		11		11					
				5					
		65		65		5		5	
		5						45	
		83		74		64		55	
		12		12		12		12	
		16		16		40		45	
						103			
		5		45				56	
		36		36					
		5		68		70		59	
		94				5			
		13		13		13		13	
520		520		520		520		520	
		10		5		33			
		27		22		20			
		5		5		6			
		39		38				14	
								5	
		22				46			
		5		5				27	
				12					
		5		16		45		45	

Course ID	Phase	Course Level	MOS	School
				RTS-M-12 FT STEWART GA
				USATC, FT. JACKSON/108TH
				Waiwa RTSM
610-63B10 Total				
610-63B30	(blank)	BNCOC	63b3	Jefferson City RTSM
				NCO ACADEMY - APG
				RTS-M, FT HOOD, TX
				RTS-M, FT INDNTWN GP PA
				RTS-M, FT MCCOY, WI
				RTS-M-01 SALINA KS
				RTS-M-02 CAMP DODGE IA
				RTS-M-03 FT DIX NJ
				RTS-M-04 FT BRAGG NC
				RTS-M-05 CAMP SHELBY MS
				RTS-M-06 CAMP ROBERT CA
				RTS-M-07 CAMP RIPLEY MN
				RTS-M-08 FT CUSTER MI
				RTS-M-09 GOWEN FIELD ID
				RTS-M-10 BLANDING FL
				RTS-M-12 FT STEWART GA
				Waiwa RTSM
610-63B30 Total				
610-63G10	(blank)	MOS	63g1	ORDNANCE SCH, APG
				RTS-M-01 SALINA KS
610-63G10 Total				
610-63S10	(blank)	MOS	63s1	Jefferson City RTSM
				RTS-M, FT HOOD, TX
				RTS-M, FT INDNTWN GP PA
				RTS-M-01 SALINA KS
				RTS-M-02 CAMP DODGE IA
				RTS-M-04 FT BRAGG NC
				RTS-M-05 CAMP SHELBY MS
				RTS-M-06 CAMP ROBERT CA
				RTS-M-08 FT CUSTER MI
				RTS-M-09 GOWEN FIELD ID
				RTS-M-12 FT STEWART GA
				USATC, FT. JACKSON/108TH
610-63S10 Total				
610-63W10	(blank)	MOS	63w1	Jefferson City RTSM
				ORDNANCE SCH, APG
				RTS-M, FT HOOD, TX

Courses Taught at Each School for the Various Options

Base Case		Nearest School		Reassign Courses Multifunctional		Reassign Courses Specialized		Consolidate Schools Specialized	
AC	RC	AC	RC	AC	RC	AC	RC	AC	RC
		19							
159		18		47				59	
		9		9		9		9	
159		159		159		159		159	
		32		5					
331		17		50		66			
		48		63		11		63	
		9		9				70	
						33			
		14		15		5		17	
		7		23					
						8			
		48		11				26	
		28		13		65		14	
		12		9		5		5	
						5		19	
		5		40		40		50	
		24		37		36		34	
		5		44		20			
		70				20		17	
		12		12		17		16	
331		331		331		331		331	
2				2		2		2	
		2							
2		2		2		2		2	
		4						46	
				7		8			
				5					
				17					
		7							
		3							
		7							
		3		5		6		6	
				5					
		14		5					
		7							
52		7		8		38			
52		52		52		52		52	
				21					
127		14		21		28		28	
		25		5		22		22	

Course ID	Phase	Course Level	MOS	School
				RTS-M, FT INDNTWN GP PA
				RTS-M, FT MCCOY, WI
				RTS-M-01 SALINA KS
				RTS-M-02 CAMP DODGE IA
				RTS-M-03 FT DIX NJ
				RTS-M-04 FT BRAGG NC
				RTS-M-06 CAMP ROBERT CA
				RTS-M-07 CAMP RIPLEY MN
				RTS-M-09 GOWEN FIELD ID
				RTS-M-10 BLANDING FL
				RTS-M-12 FT STEWART GA
610-63W10 Total				
610-ASIH8 (63B/S)	(blank)	ASI	63B/S	Jefferson City RTSM
				RTS-M, FT HOOD, TX
				RTS-M, FT INDNTWN GP PA
				RTS-M, FT MCCOY, WI
				RTS-M-02 CAMP DODGE IA
				RTS-M-04 FT BRAGG NC
				RTS-M-06 CAMP ROBERT CA
				RTS-M-10 BLANDING FL
				USATC, FT. JACKSON/1O8TH
				Waiwa RTSM
610-ASIH8 (63B/S) Total				
611-63D10	(blank)	MOS	63d1	ORDNANCE SCH, APG
				RTS-M-01 SALINA KS
				RTS-M-05 CAMP SHELBY MS
				RTS-M-07 CAMP RIPLEY MN
611-63D10 Total				
611-63D30	(blank)	BNCOC	63d3	NCO ACADEMY - APG
				RTS-M, FT HOOD, TX
				RTS-M, FT MCCOY, WI
				RTS-M-01 SALINA KS
				RTS-M-07 CAMP RIPLEY MN
				RTS-M-12 FT STEWART GA
611-63D30 Total				
611-63D30 (45D)	(blank)	BNCOC	63d3	NCO ACADEMY - APG
				RTS-M-01 SALINA KS
				RTS-M-12 FT STEWART GA
611-63D30 (45D) Total				
611-63E10	(blank)	MOS	63E10	ARMOR SCH, FT KNOX
				RTS-M, FT INDNTWN GP PA

Courses Taught at Each School for the Various Options

Base Case		Nearest School		Reassign Courses Multifunctional		Reassign Courses Specialized		Consolidate Schools Specialized	
AC	RC	AC	RC	AC	RC	AC	RC	AC	RC
		7		7					
				5		46			
		24							
				17					
		7						19	
		8		8		15		15	
				14				27	
		11		17					
				12		16		16	
		31							
127		127		127		127		127	
				38				39	
				62		72		72	
				40		40		40	
		249							
				37		39			
				23					
		32		32		32		32	
				49					
286						98		98	
		5		5		5		5	
286		286		286		286		286	
8									
				8					
		8							
						8		8	
8		8		8		8		8	
34									
		23		8					
		6		5					
				16		19		19	
		5		5		15		15	
34		34		34		34		34	
7									
		7		7		7		7	
7		7		7		7		7	
34		8		13					
				5				9	

Course ID	Phase	Course Level	MOS	School
				RTS-M-02 CAMP DODGE IA
				RTS-M-06 CAMP ROBERT CA
				RTS-M-07 CAMP RIPLEY MN
				RTS-M-08 FT CUSTER MI
				RTS-M-09 GOWEN FIELD ID
				RTS-M-12 FT STEWART GA
611-63E10 Total				
611-63E30	(blank)	BNCOC	63E30	NCO ACADEMY - FT KNOX
				RTS-M, FT INDNTWN GP PA
				RTS-M-01 SALINA KS
				RTS-M-02 CAMP DODGE IA
				RTS-M-06 CAMP ROBERT CA
				RTS-M-07 CAMP RIPLEY MN
				RTS-M-08 FT CUSTER MI
				RTS-M-09 GOWEN FIELD ID
				RTS-M-12 FT STEWART GA
611-63E30 Total				
611-63H10	(blank)	MOS	63h1	Jefferson City RTSM
				ORDNANCE SCH, APG
				RTS-M, FT INDNTWN GP PA
				RTS-M, FT MCCOY, WI
				RTS-M-01 SALINA KS
				RTS-M-02 CAMP DODGE IA
				RTS-M-03 FT DIX NJ
				RTS-M-04 FT BRAGG NC
				RTS-M-05 CAMP SHELBY MS
				RTS-M-06 CAMP ROBERT CA
				RTS-M-08 FT CUSTER MI
				RTS-M-09 GOWEN FIELD ID
				RTS-M-12 FT STEWART GA
611-63H10 Total				
611-63H30	(blank)	BNCOC	63h3	Jefferson City RTSM
				NCO ACADEMY - APG
				RTS-M-02 CAMP DODGE IA
				RTS-M-05 CAMP SHELBY MS
611-63H30 Total				
611-63T10	(blank)	MOS	63t1	ARMOR SCH, FT KNOX
				RTS-M, FT HOOD, TX
				RTS-M, FT INDNTWN GP PA
				RTS-M-01 SALINA KS
				RTS-M-03 FT DIX NJ

Base Case		Nearest School		Reassign Courses Multifunctional		Reassign Courses Specialized		Consolidate Schools Specialized	
AC	RC	AC	RC	AC	RC	AC	RC	AC	RC
				6		25		25	
		5		5					
		6							
		5		5					
		10				9			
34		34		34		34		34	
31		7		11					
		5		5				7	
						15			
		5		5					
				5		11		7	
								17	
		5		5					
		9				5			
31		31		31		31		31	
		12							
83		48		7					
		5							
				48					
								19	
						42			
		18				5			
								44	
				5		5			
				5		31			
				5				20	
				13					
83		83		83		83		83	
				39		21		5	
119		119		47		42		57	
						35		57	
				33		21			
119		119		119		119		119	
84									
				5		25		24	
		24				8			
				16				6	
		6							

Courses Taught at Each School for the Various Options

Course ID	Phase	Course Level	MOS	School
				RTS-M-04 FT BRAGG NC
				RTS-M-05 CAMP SHELBY MS
				RTS-M-06 CAMP ROBERT CA
				RTS-M-07 CAMP RIPLEY MN
				RTS-M-08 FT CUSTER MI
				RTS-M-09 GOWEN FIELD ID
				RTS-M-12 FT STEWART GA
611-63T10 Total				
611-63T30	(blank)	BNCOC	63t3	NCO ACADEMY - FT KNOX
				RTS-M, FT HOOD, TX
				RTS-M-01 SALINA KS
				RTS-M-03 FT DIX NJ
				RTS-M-04 FT BRAGG NC
				RTS-M-05 CAMP SHELBY MS
				RTS-M-06 CAMP ROBERT CA
				RTS-M-07 CAMP RIPLEY MN
				RTS-M-09 GOWEN FIELD ID
				RTS-M-12 FT STEWART GA
611-63T30 Total				
611-63Y10	(blank)	MOS	63y1	ORDNANCE SCH, APG
				RTS-M, FT HOOD, TX
				RTS-M-01 SALINA KS
				RTS-M-03 FT DIX NJ
				RTS-M-08 FT CUSTER MI
				RTS-M-09 GOWEN FIELD ID
611-63Y10 Total				
611-ASIH8 (63D/H/W/Y)	(blank)	ASI	63DH WYasi	ORDNANCE SCH, APG
611-ASIH8 (63D/H/W/Y) Total				
611-ASIH8 (63E)	(blank)	ASI	63E	ARMOR SCH, FT KNOX
611-ASIH8 (63E) Total				
611-ASIH8 (63T)	(blank)	ASI	63T	ARMOR SCH, FT KNOX
611-ASIH8 (63T) Total				
612-62B10	(blank)	MOS	62b1	RTS-M, FT HOOD, TX
				RTS-M, FT INDNTWN GP PA
				RTS-M, FT MCCOY, WI
				RTS-M-04 FT BRAGG NC
				RTS-M-05 CAMP SHELBY MS
				RTS-M-06 CAMP ROBERT CA
				RTS-M-07 CAMP RIPLEY MN

Courses Taught at Each School for the Various Options

Base Case		Nearest School		Reassign Courses Multifunctional		Reassign Courses Specialized		Consolidate Schools Specialized	
AC	RC	AC	RC	AC	RC	AC	RC	AC	RC
				5		23		5	
		13		6		6		27	
		5		5				5	
		14		29		17			
		22		7		5			
				11				17	
84		84		84		84		84	
29		5		14					
		8							
								8	
						8			
								15	
						15			
		6		5		6		6	
		5		5					
				5					
		5							
29		29		29		29		29	
25				10					
		13		5					
				5					
						25			
				5				25	
		12							
25		25		25		25		25	
155		155		155		155		155	
155		155		155		155		155	
67		67		67		67		67	
67		67		67		67		67	
102		102		102		102		102	
102		102		102		102		102	
						25			
				5		8		7	
		12							
				8		25		27	
				5					
		12		5					
		14						26	

Course ID	Phase	Course Level	MOS	School
				RTS-M-08 FT CUSTER MI
				RTS-M-09 GOWEN FIELD ID
				USATC, FT. WOOD/98TH DIV
612-62B10 Total				
612-62B30	(blank)	BNCOC	62b3	Ft. Devens RTSM
				Jefferson City RTSM
				NCO ACADEMY - FT L. WOOD
				RTS-M, FT HOOD, TX
				RTS-M, FT INDNTWN GP PA
				RTS-M, FT MCCOY, WI
				RTS-M-04 FT BRAGG NC
				RTS-M-06 CAMP ROBERT CA
				RTS-M-08 FT CUSTER MI
				RTS-M-09 GOWEN FIELD ID
				RTS-M-12 FT STEWART GA
612-62B30 Total				
641-45B10	(blank)	MOS	45b1	ORDNANCE SCH, APG
				RTS-M, FT HOOD, TX
				RTS-M-01 SALINA KS
				RTS-M-02 CAMP DODGE IA
				RTS-M-09 GOWEN FIELD ID
				RTS-M-10 BLANDING FL
641-45B10 Total				
642-45D10	(blank)	MOS	45d1	FLD ARTILLERY SCH, SILL
				RTS-M-07 CAMP RIPLEY MN
642-45D10 Total				
643-45E10	(blank)	MOS	45E10	ARMOR SCH, FT KNOX
				RTS-M-06 CAMP ROBERT CA
				RTS-M-07 CAMP RIPLEY MN
				RTS-M-09 GOWEN FIELD ID
643-45E10 Total				
643-45K10	(blank)	MOS	45k1	ORDNANCE SCH, APG
				RTS-M-02 CAMP DODGE IA
				RTS-M-04 FT BRAGG NC
				RTS-M-06 CAMP ROBERT CA
				RTS-M-09 GOWEN FIELD ID
643-45K10 Total				
643-45K30	(blank)	BNCOC	45k3	NCO ACADEMY - APG
				RTS-M, FT INDNTWN GP PA
				RTS-M-02 CAMP DODGE IA
				RTS-M-04 FT BRAGG NC

Courses Taught at Each School for the Various Options

Base Case		Nearest School		Reassign Courses Multifunctional		Reassign Courses Specialized		Consolidate Schools Specialized	
AC	RC	AC	RC	AC	RC	AC	RC	AC	RC
				5					
		28		5		8		6	
66				33					
66		66		66		66		66	
		3							
		11							
45				12					
						8			
				5					
		4							
				7					
		27		5					
				5					
				6		12		14	
				5		25		31	
45		45		45		45		45	
24				9					
		11							
						24		24	
		1		5					
		12		5					
				5					
24		24		24		24		24	
14		14		14					
						14		14	
14		14		14		14		14	
8						8			
				8					
								8	
		8							
8		8		8		8		8	
13				8					
				5					
						8		8	
		13							
						5		5	
13		13		13		13		13	
48		35		10					
		3							
				19		12		12	
		10				6		6	

Course ID	Phase	Course Level	MOS	School
				RTS-M-09 GOWEN FIELD ID
				RTS-M-10 BLANDING FL
				RTS-M-12 FT STEWART GA
643-45K30 Total				
643-45T10	(blank)	MOS	45t1	ARMOR SCH, FT KNOX
643-45T10 Total				
662-52C10	(blank)	MOS	52c1	Jefferson City RTSM
				ORDNANCE SCH, APG
				RTS-M-01 SALINA KS
				RTS-M-02 CAMP DODGE IA
				RTS-M-12 FT STEWART GA
662-52C10 Total				
662-52C30	(blank)	BNCOC	52c3	Jefferson City RTSM
				NCO ACADEMY - APG
				RTS-M-01 SALINA KS
				RTS-M-02 CAMP DODGE IA
				RTS-M-12 FT STEWART GA
662-52C30 Total				
662-52C30 (63J)	(blank)	BNCOC	52c3	Jefferson City RTSM
				NCO ACADEMY - APG
				RTS-M-01 SALINA KS
				RTS-M-02 CAMP DODGE IA
				RTS-M-12 FT STEWART GA
662-52C30 (63J) Total				
662-52D10	(blank)	MOS	52d1	ORDNANCE SCH, APG
				RTS-M, FT HOOD, TX
				RTS-M, FT MCCOY, WI
				RTS-M-02 CAMP DODGE IA
				RTS-M-03 FT DIX NJ
				RTS-M-04 FT BRAGG NC
				RTS-M-05 CAMP SHELBY MS
				RTS-M-06 CAMP ROBERT CA
				RTS-M-08 FT CUSTER MI
				RTS-M-10 BLANDING FL
662-52D10 Total				
662-52D30	(blank)	BNCOC	52d3	Jefferson City RTSM
				NCO ACADEMY - APG
				RTS-M, FT HOOD, TX
				RTS-M, FT MCCOY, WI
				RTS-M-02 CAMP DODGE IA
				RTS-M-03 FT DIX NJ

Courses Taught at Each School for the Various Options

Base Case		Nearest School		Reassign Courses Multifunctional		Reassign Courses Specialized		Consolidate Schools Specialized	
AC	RC	AC	RC	AC	RC	AC	RC	AC	RC
				13		9		9	
				6		14		14	
						6		6	
48		48		48		47		47	
6		6		6		6		6	
6		6		6		6		6	
		12				19		19	
				8					
19									
		7		5					
				6					
19		19		19		19		19	
				5					
25		5		5					
				5		15		16	
				5				9	
		20		5		10			
25		25		25		25		25	
				5		12			
24				7				24	
						12			
				7					
		24		5					
24		24		24		24		24	
84				35		5			
						12		25	
		20				17			
		38		8					
				5				25	
						18			
				7				17	
		26		6		8		6	
				5					
				18		25		11	
84		84		84		85		84	
		5						5	
47		5		6					
						5			
				16					
		5							
				5					

Course ID	Phase	Course Level	MOS	School
				RTS-M-04 FT BRAGG NC
				RTS-M-05 CAMP SHELBY MS
				RTS-M-06 CAMP ROBERT CA
				RTS-M-08 FT CUSTER MI
				RTS-M-10 BLANDING FL
				RTS-M-12 FT STEWART GA
662-52D30 Total				
662-52F10	(blank)	MOS	52f1	ORDNANCE SCH, APG
662-52F10 Total				
662-52F30	(blank)	BNCOC	52f3	NCO ACADEMY - APG
662-52F30 Total				
662-ASIC9	(blank)	ASI	52D	ORDNANCE SCH, APG
662-ASIC9 Total				
690-63J10	(blank)	MOS	63j1	ORDNANCE SCH, APG
				RTS-M, FT INDNTWN GP PA
				RTS-M-02 CAMP DODGE IA
				RTS-M-12 FT STEWART GA
690-63J10 Total				
702-44E10	(blank)	MOS	44E10	ORDNANCE SCH, APG
				RTS-M-01 SALINA KS
702-44E10 Total				
702-44E30	(blank)	BNCOC	44E30	NCO ACADEMY - APG
				RTS-M-01 SALINA KS
702-44E30 Total				
702-44E30 (44B)	(blank)	BNCOC	44E30 (44B)	NCO ACADEMY - APG
				RTS-M-01 SALINA KS
702-44E30 (44B) Total				
704-44B10	(blank)	MOS	44b1	ORDNANCE SCH, APG
				RTS-M-01 SALINA KS
				RTS-M-12 FT STEWART GA
704-44B10 Total				
M1070 NET (63H1)	(blank)	New Eqpt	63H	ORDNANCE SCH, APG
				RTS-M, FT MCCOY, WI
M1070 NET (63H1) Total				
M1074 NET (63H)	(blank)	New Eqpt	63H	ORDNANCE SCH, APG
				RTS-M, FT HOOD, TX
M1074 NET (63H) Total				
M109A6 (45D)	(blank)	New Eqpt	45D	FLD ARTILLERY SCH, SILL
				RTS-M, FT HOOD, TX
M109A6 (45D) Total				
M109A6 (63D)	(blank)	New Eqpt	63D	ORDNANCE SCH, APG

Courses Taught at Each School for the Various Options

Base Case		Nearest School		Reassign Courses Multifunctional		Reassign Courses Specialized		Consolidate Schools Specialized	
AC	RC	AC	RC	AC	RC	AC	RC	AC	RC
		6							
		8				12			
		8		8		10		12	
				5		19		19	
		5		7				11	
		5							
47		47		47		46		47	
5		5		5		5		5	
5		5		5		5		5	
8		8		8		8		8	
8		8		8		8		8	
6		6		6		6		6	
6		6		6		6		6	
46		41		19					
		5		5		24		16	
				14		12		8	
				8		10		22	
46		46		46		46		46	
1						1		1	
		1		1					
1		1		1		1		1	
17		17		9					
				8		17		17	
17		17		17		17		17	
21		21		7					
				14		21		21	
21		21		21		21		21	
28				11					
		28		12		11		7	
				5		17		21	
28		28		28		28		28	
									4
	4		4		4		4		
	4		4		4		4		4
			5		5		53		53
	53		48		48				
	53		53		53		53		53
					2		2		2
	2		2						
	2		2		2		2		2
							1		1

Course ID	Phase	Course Level	MOS	School
				RTS-M, FT HOOD, TX
M109A6 (63D) Total				
TAMMS	(blank)	Other	?	NCO ACADEMY - FT LEE
				RTS-M-02 CAMP DODGE IA
				RTS-M-03 FT DIX NJ
				RTS-M-05 CAMP SHELBY MS
				Waiwa RTSM
TAMMS Total				
ULLS	(blank)	Other	?	Ft. Devens RTSM
				NCO ACADEMY - FT LEE
				RTS-M, FT HOOD, TX
				RTS-M, FT MCCOY, WI
ULLS Total				
ULLS-G (S)	(blank)	Sustainment	?	Jefferson City RTSM
				NCO ACADEMY - FT LEE
				RTS-M-02 CAMP DODGE IA
				RTS-M-03 FT DIX NJ
				RTS-M-04 FT BRAGG NC
				RTS-M-05 CAMP SHELBY MS
				RTS-M-06 CAMP ROBERT CA
				RTS-M-07 CAMP RIPLEY MN
				RTS-M-09 GOWEN FIELD ID
				RTS-M-10 BLANDING FL
				RTS-M-12 FT STEWART GA
				Waiwa RTSM
ULLS-G (S) Total				
ULLS-G (SCP-05)	(blank)	Other	?	NCO ACADEMY - FT LEE
				RTS-M, FT HOOD, TX
ULLS-G (SCP-05) Total				

Courses Taught at Each School for the Various Options

Base Case		Nearest School		Reassign Courses Multifunctional		Reassign Courses Specialized		Consolidate Schools Specialized	
AC	RC	AC	RC	AC	RC	AC	RC	AC	RC
	1		1		1				
	1		1		1		1		1
			25		23				
	8		12		5				
	22								
	7				5		33		33
	1		1		5		5		5
	38		38		38		38		38
	44				41				
			66		5		79		79
	22				22				
	13		13		11				
	79		79		79		79		79
	19		19		27				
			25		22				
	101		72		41				
	57		57		60				
	36		38		38		38		38
	6		6		12				
	20		20		20				
	17		19		23				
	55		55		72		330		330
	81		81		84		98		98
	38		41		34				
	36		33		33				
	466		466		466		466		466
							57		57
	57		57		57				
	57		57		57		57		57

Appendix C
SCHOOLS OFFERING SPECIFIC COURSES FOR THE VARIOUS OPTIONS

This appendix shows the various AC schools and the RTS-Ms where each course is offered for the various options described in the body of the report. For each course, the tables show the course identifier, the specific phases for the RC courses, the course level (MOS reclassification, Advance Skill Identifier (ASI), BNCOC, ANCOC, etc.), the MOS appropriate for the course, the schools where the course is offered, and the number of AC and RC students at each school and in total for the base case (i.e., as reflected in the fiscal year 1996 ATRRS database) and for each of the three options examined during the analyses.

School	Course ID	Phase	Course Level	MOS
ARMOR SCH, FT KNOX	091-45E10	(blank)	MOS	45E10
	091-63E10	1	MOS	63E10
		2	MOS	63E10
		(blank)	MOS	63E10
	091-63T10	1	MOS	63t1
		2	MOS	63t1
	611-63E10	(blank)	MOS	63E10
	611-63T10	(blank)	MOS	63t1
	611-ASIH8 (63E)	(blank)	ASI	63E
	611-ASIH8 (63T)	(blank)	ASI	63T
	643-45E10	(blank)	MOS	45E10
	643-45T10	(blank)	MOS	45t1
ARMOR SCH, FT KNOX Total				
FLD ARTILLERY SCH, SILL	091-45D10	1	MOS	45d1
		2	MOS	45d1
	642-45D10	(blank)	MOS	45d1
	M109A6 (45D)	(blank)	New Eqpt	45D
FLD ARTILLERY SCH, SILL Total				
Ft. Devens RTSM	052-62B10	(blank)	MOS	62b1
	052-62B30	2	BNCOC	62b3
	091-52D10	2	MOS	52d1
	091-63H30	2	BNCOC	63h3
	6-63-C42	(blank)	ANCOC	63-all
	612-62B30	(blank)	BNCOC	62b3
	ULLS	(blank)	Other	?
Ft. Devens RTSM Total				
Jefferson City RTSM	052-62B10	(blank)	MOS	62b1
	052-62B30	2	BNCOC	62b3
	052-62B40	2	ANCOC	62B4
	091-52C10	1	MOS	52c1
		2	MOS	52c1
	091-52C30	2	BNCOC	52c3
	091-52D10	2	MOS	52d1
	091-52D30	2	BNCOC	52d3
	091-63B/S/W10H8	(blank)	ASI	63BSWasi
	091-63B10	1	MOS	63b1
		2	MOS	63b1
	091-63B30	2	BNCOC	63b3
	091-63B40	2	ANCOC	63B4
	091-63H10	1	MOS	63h1
		2	MOS	63h1

Schools Offering Specific Courses for the Various Options

Base Case		Nearest School		Reassign Courses Multifunctional		Reassign Courses Specialized		Consolidate Schools Specialized	
AC	RC	AC	RC	AC	RC	AC	RC	AC	RC
			25		13				
					5				
			48		5				
			39		23				
					5				
			5		5				
34		8		13					
84									
67		67		67		67		67	
102		102		102		102		102	
8						8			
6		6		6		6		6	
301		183		117	188	56	183		175
			11		5		11		11
			20		6		20		20
14		14		14					
					2		2		2
14		14	31	14	13		33		33
	11		9		9				
	3		3		9				
	6		6		6				
	4		4		5				
			5		5		13		
			3						
	44				41				
	68	8	22	5	70	13			
	20		5						
	6		4						
	16								
	9		12		28				
	10		12		19				
	6		16		5		16		16
	6		24		5				
	7		7		7		28		
	7		12		24				
	4		21		11				
	28		51		36				36
	19		32		25				
	21		28		19				
	5		9		42				
	4		12		6		6		

School	Course ID	Phase	Course Level	MOS
	091-63H30-IIA	2	BNCOC	63h3
	091-63H30-IIB	2	BNCOC	63h3
	091-63S10	1	MOS	63s1
		2	MOS	63s1
	091-63W10	1	MOS	63w1
		2	MOS	63w1
	101-92A10	1	MOS	92a1
		2	MOS	92a1
	101-92A30	2	BNCOC	92a3
	101-92A40	2	ANCOC	92a4
	551-92A10	(blank)	MOS	92a1
	551-92A30	(blank)	BNCOC	92a3
	551-92A40	(blank)	ANCOC	92a4
	6-63-C42	(blank)	ANCOC	63-all
	610-63B10	(blank)	MOS	63b1
	610-63B30	(blank)	BNCOC	63b3
	610-63S10	(blank)	MOS	63s1
	610-63W10	(blank)	MOS	63w1
	610-ASIH8 (63B/S)	(blank)	ASI	63B/S
	611-63H10	(blank)	MOS	63h1
	611-63H30	(blank)	BNCOC	63h3
	612-62B30	(blank)	BNCOC	62b3
	662-52C10	(blank)	MOS	52c1
	662-52C30	(blank)	BNCOC	52c3
	662-52C30 (63J)	(blank)	BNCOC	52c3
	662-52D30	(blank)	BNCOC	52d3
	ULLS-G (S)	(blank)	Sustainment	?
Jefferson City RTSM Total				
NCO ACADEMY - APG	091-44E30	2	BNCOC	44E30
		3	BNCOC	44E30
	091-45K30	2	BNCOC	45k3
		3	BNCOC	45k3
	091-45K40	2	ANCOC	45K4
	091-52C30	2	BNCOC	52c3
	091-52D30	2	BNCOC	52d3
	091-63B30	2	BNCOC	63b3
	091-63B40	2	ANCOC	63B4
	091-63D30 TRK II	2	BNCOC	63d3
	091-63D40	2	ANCOC	63D4
	091-63H30-IIA	2	BNCOC	63h3
	091-63H30-IIB	2	BNCOC	63h3

Schools Offering Specific Courses for the Various Options 93

Base Case		Nearest School		Reassign Courses Multifunctional		Reassign Courses Specialized		Consolidate Schools Specialized	
AC	RC	AC	RC	AC	RC	AC	RC	AC	RC
5				5					
12		12		10					40
6		6		5					53
12		11		11					
25		38		23					68
17		38		20		26			20
31		40		54		64			65
24		40		29		47			36
21				9		47			47
15		20		5					
		14		5		14		14	
		40		8		70		51	
				26					
		26		26		44		42	
		10		5		33			
		32		5					
		4						46	
				21					
				38				39	
		12							
				39		21		5	
		11							
		12				19		19	
				5					
				5		12			
		5						5	
19		19		27					
355		**166**	**469**	**183**	**425**	**213**	**234**	**221**	**381**
							8		8
				8		8		8	
				6					
				2					
				6					
				6					
				13		13			
				18		13			
						5		12		12
						8		8		8
				12		10			
						13		40		

School	Course ID	Phase	Course Level	MOS
	091-63H40	2	ANCOC	63H4
	6-63-C42	(blank)	ANCOC	63-all
	610-63B30	(blank)	BNCOC	63b3
	611-63D30	(blank)	BNCOC	63d3
	611-63D30 (45D)	(blank)	BNCOC	63d3
	611-63H30	(blank)	BNCOC	63h3
	643-45K30	(blank)	BNCOC	45k3
	662-52C30	(blank)	BNCOC	52c3
	662-52C30 (63J)	(blank)	BNCOC	52c3
	662-52D30	(blank)	BNCOC	52d3
	662-52F30	(blank)	BNCOC	52f3
	702-44E30	(blank)	BNCOC	44E30
	702-44E30 (44B)	(blank)	BNCOC	44E30 (44B)
NCO ACADEMY - APG Total				
NCO ACADEMY - FT KNOX	091-63E30	2	BNCOC	63E30
	091-63E40	2	ANCOC	63E40
	091-63T30	2	BNCOC	63t3
	091-63T40	2	ANCOC	63T4
	611-63E30	(blank)	BNCOC	63E30
	611-63T30	(blank)	BNCOC	63t3
NCO ACADEMY - FT KNOX Total				
NCO ACADEMY - FT L. WOOD	052-62B30	2	BNCOC	62b3
	612-62B30	(blank)	BNCOC	62b3
NCO ACADEMY - FT L. WOOD Total				
NCO ACADEMY - FT LEE	101-92A30	2	BNCOC	92a3
	101-92A40	2	ANCOC	92a4
	551-92A30	(blank)	BNCOC	92a3
	551-92A40	(blank)	ANCOC	92a4
	TAMMS	(blank)	Other	?
	ULLS	(blank)	Other	?
	ULLS-G (S)	(blank)	Sustainment	?
	ULLS-G (SCP-05)	(blank)	Other	?
NCO ACADEMY - FT LEE Total				
ORDNANCE SCH, APG	091-44B10	1	MOS	44b1
		2	MOS	44b1
		3	MOS	44b1
	091-44E10	1	MOS	44E10
	091-45B10	1	MOS	45b1
		2	MOS	45b1
		(blank)	MOS	45b1
	091-45K10	1	MOS	45k1

Schools Offering Specific Courses for the Various Options

Base Case		Nearest School		Reassign Courses Multifunctional		Reassign Courses Specialized		Consolidate Schools Specialized	
AC	RC	AC	RC	AC	RC	AC	RC	AC	RC
			5		5				
520		60		60		67		104	
331		17		50		66			
34									
7									
119		119		47		42		57	
48		35		10					
25		5		5					
24				7				24	
47		5		6					
8		8		8		8		8	
17		17		9					
21		21		7					
1201		287	58	209	93	183	76	193	36
			12				12		12
					13				
					7				
			11		5		11		11
31		7			11				
29		5			14				
60		12	23	25	25		23		23
			3		9				
45				12					
45			3	12	9				
			9		5				
			2		5				
252						84		99	
135				74		80		81	
			25		23				
			66		5		79		79
			25		22				
							57		57
387			127	74	60	164	136	180	136
			45		7		45		45
					5				
							9		9
					3		3		3
					5		6		6
					5		12		12
			16		5		20		20
			7		8		8		8

School	Course ID	Phase	Course Level	MOS
		2	MOS	45k1
		3	MOS	45k1
	091-45K10 (T)	(blank)	Transition	45K
	091-52C10	1	MOS	52c1
		2	MOS	52c1
	091-52D10	1	MOS	52d1
		2	MOS	52d1
		(blank)	MOS	52d1
	091-52X40	2	ANCOC	52X4
	091-63B/S/W10H8	(blank)	ASI	63BSWasi
	091-63D10	1	MOS	63d1
		2	MOS	63d1
		(blank)	MOS	63d1
	091-63G10	2	MOS	63g1
	091-63H10	(blank)	MOS	63h1
	091-63H10 (GS) BFV	2	MOS	63h1
	091-63H10 (GS) M1	(blank)	MOS	63h1
	091-63H10 (T)	(blank)	Transition	63H
	091-63H10L8	(blank)	ASI	63H
	091-63J10	1	MOS	63j1
		2	MOS	63j1
	091-63W10	1	MOS	63w1
		2	MOS	63w1
		(blank)	MOS	63w1
	091-63W10 (GS) HEMTT	1	MOS	63w1
	091-63W10 (GS) HMMWV	1	MOS	63w1
		2	MOS	63w1
	091-63W10 (GS) M939	1	MOS	63w1
	091-63Y10	1	MOS	63y1
		2	MOS	63y1
	091-ASIL8 (45K)	1	ASI	45Kasi
	113-45G10	(blank)	MOS	45g1
	610-63G10	(blank)	MOS	63g1
	610-63W10	(blank)	MOS	63w1
	611-63D10	(blank)	MOS	63d1
	611-63H10	(blank)	MOS	63h1
	611-63Y10	(blank)	MOS	63y1
	611-ASIH8 (63D/H/W/Y)	(blank)	ASI	63DHWYasi

Schools Offering Specific Courses for the Various Options

Base Case		Nearest School		Reassign Courses Multifunctional		Reassign Courses Specialized		Consolidate Schools Specialized	
AC	RC	AC	RC	AC	RC	AC	RC	AC	RC
						12		12	
		26		5					
						14		14	
		43		17					
		43		6		55		13	
		21		5					
		2		5					
				5		11		11	
		7		7		12			
		11		11					
		5		5		11		11	
		43		5					
		20		8		20		20	
				5		11		11	
		23		11		23		23	
				5		5		5	
		8		8		8		8	
		11							
				6		6		6	
		50		5					
		26		14		26			
		36		5				37	
		12		5					
								21	
		12		5		12		12	
				3		3		3	
				3		3		3	
				8		8		8	
				13					
		9		9					
						6		6	
40		40		40		40		40	
2				2		2		2	
127		14		21		28		28	
8									
83		48		7					
25				10					
155		155		155		155		155	

School	Course ID	Phase	Course Level	MOS
	641-45B10	(blank)	MOS	45b1
	643-45K10	(blank)	MOS	45k1
	662-52C10	(blank)	MOS	52c1
	662-52D10	(blank)	MOS	52d1
	662-52F10	(blank)	MOS	52f1
	662-ASIC9	(blank)	ASI	52D
	690-63J10	(blank)	MOS	63j1
	702-44E10	(blank)	MOS	44E10
	704-44B10	(blank)	MOS	44b1
	M1070 NET (63H1)	(blank)	New Eqpt	63H
	M1074 NET (63H)	(blank)	New Eqpt	63H
	M109A6 (63D)	(blank)	New Eqpt	63D
ORDNANCE SCH, APG Total				
QUARTERMASTER,FT LEE	101-92A10	1	MOS	92a1
		2	MOS	92a1
	551-92A10	(blank)	MOS	92a1
QUARTERMASTER,FT LEE Total				
RTS-M, FT HOOD, TX	052-62B10	(blank)	MOS	62b1
	052-62B30	2	BNCOC	62b3
	091-45B10	1	MOS	45b1
		2	MOS	45b1
	091-45K10	1	MOS	45k1
	091-45K30	2	BNCOC	45k3
		3	BNCOC	45k3
	091-52D10	1	MOS	52d1
		2	MOS	52d1
	091-52D30	2	BNCOC	52d3
	091-62B40	2	ANCOC	62B4
	091-63B/S/W10H8	(blank)	ASI	63BSWasi
	091-63B10	1	MOS	63b1
		2	MOS	63b1
	091-63B30	2	BNCOC	63b3
	091-63B40	2	ANCOC	63B4
	091-63D/E/H/N/T/Y10H8	(blank)	ASI	63DEHNTYasi
	091-63D10	2	MOS	63d1
	091-63H10	1	MOS	63h1
		2	MOS	63h1
	091-63H30	2	BNCOC	63h3
	091-63H40	2	ANCOC	63H4
	091-63S10	2	MOS	63s1

Schools Offering Specific Courses for the Various Options

Base Case		Nearest School		Reassign Courses Multifunctional		Reassign Courses Specialized		Consolidate Schools Specialized	
AC	RC	AC	RC	AC	RC	AC	RC	AC	RC
24				9					
13				8					
19				8					
84				35		5			
5		5		5		5		5	
6		6		6		6		6	
46		41		19					
1						1		1	
28				11					
									4
			5		5		53		53
							1		1
666		309	481	336	227	242	403	237	385
					7				
			10		9				
231		28				100		100	
231		28	10		16	100		100	
	11		7		5		17		
	8		3		5				
	3		3		5				
	3								
	4		5		5				
	3		3						
	7				7				
	4		15		6		14		
	5		8		7				
	10		13		13		30		
	4		4		5				
	11		10		5		10		18
	2		5		5		5		5
	35		35		26		56		26
	14		17		9		17		17
	8		8		8				15
	10		5		5		10		10
	24		24		54				
	14		8		5		5		5
	19		21		31		32		52
	23		23		22		46		16
	10		10		8				
	5		3		5				6

100 Consolidating Active and Reserve Component Training Infrastructure

School	Course ID	Phase	Course Level	MOS
	091-63T10	2	MOS	63t1
	091-63T30	2	BNCOC	63t3
	091-63W10	2	MOS	63w1
	091-63Y10	2	MOS	63y1
	6-63-C42	(blank)	ANCOC	63-all
	610-63B10	(blank)	MOS	63b1
	610-63B30	(blank)	BNCOC	63b3
	610-63S10	(blank)	MOS	63s1
	610-63W10	(blank)	MOS	63w1
	610-ASIH8 (63B/S)	(blank)	ASI	63B/S
	611-63D30	(blank)	BNCOC	63d3
	611-63T10	(blank)	MOS	63t1
	611-63T30	(blank)	BNCOC	63t3
	611-63Y10	(blank)	MOS	63y1
	612-62B10	(blank)	MOS	62b1
	612-62B30	(blank)	BNCOC	62b3
	641-45B10	(blank)	MOS	45b1
	662-52D10	(blank)	MOS	52d1
	662-52D30	(blank)	BNCOC	52d3
	M1074 NET (63H)	(blank)	New Eqpt	63H
	M109A6 (45D)	(blank)	New Eqpt	45D
	M109A6 (63D)	(blank)	New Eqpt	63D
	ULLS	(blank)	Other	?
	ULLS-G (SCP-05)	(blank)	Other	?
RTS-M, FT HOOD, TX Total				
RTS-M, FT INDNTWN GP PA	052-62B10	(blank)	MOS	62b1
	052-62B30	2	BNCOC	62b3
	091-45K10	2	MOS	45k1
	091-52X40	2	ANCOC	52X4
	091-63B/S/W10H8	(blank)	ASI	63BSWasi
	091-63B10	2	MOS	63b1
	091-63B30	2	BNCOC	63b3
	091-63B40	2	ANCOC	63B4
	091-63D/E/H/N/T/Y10H8	(blank)	ASI	63DEHNTYasi
	091-63D10	2	MOS	63d1
	091-63E10	2	MOS	63E10
	091-63E30	2	BNCOC	63E30
	091-63H10	2	MOS	63h1
	091-63H30	2	BNCOC	63h3
	091-63H40	2	ANCOC	63H4

Schools Offering Specific Courses for the Various Options

Base Case		Nearest School		Reassign Courses Multifunctional		Reassign Courses Specialized		Consolidate Schools Specialized	
AC	RC	AC	RC	AC	RC	AC	RC	AC	RC
	5		5		5		7		12
	5								
	13		25		9		9		13
	7		5		5				
		84		84		84		84	
		27		22		20			
		48		63		11		63	
				7		8			
		25		5		22		22	
				62		72		72	
		23		8					
				5		25		24	
		8							
		13		5					
						25			
						8			
		11							
						12		25	
						5			
53		48		48					
2		2							
1		1		1					
22				22					
57		57		57					
402	239	373	261	388	292	258	290	195	
14		32		30		73		47	
2		12		9		28		52	
6		25		10					
12		5		5				12	
7		8		8					
18		24		33					
9		27		17					
6		12		12		39		39	
10		10		21					
6		15		25					
7		8		7		42		42	
6				5					
25		25		33		50		38	
5		5		5		33		63	
5		10		10		24		35	

102 Consolidating Active and Reserve Component Training Infrastructure

School	Course ID	Phase	Course Level	MOS
	091-63J10	1	MOS	63j1
		2	MOS	63j1
	091-63S10	2	MOS	63s1
	091-63T10	2	MOS	63t1
	091-63W10	1	MOS	63w1
		2	MOS	63w1
	6-63-C42	(blank)	ANCOC	63-all
	610-63B10	(blank)	MOS	63b1
	610-63B30	(blank)	BNCOC	63b3
	610-63S10	(blank)	MOS	63s1
	610-63W10	(blank)	MOS	63w1
	610-ASIH8 (63B/S)	(blank)	ASI	63B/S
	611-63E10	(blank)	MOS	63E10
	611-63E30	(blank)	BNCOC	63E30
	611-63H10	(blank)	MOS	63h1
	611-63T10	(blank)	MOS	63t1
	612-62B10	(blank)	MOS	62b1
	612-62B30	(blank)	BNCOC	62b3
	643-45K30	(blank)	BNCOC	45k3
	690-63J10	(blank)	MOS	63j1
RTS-M, FT INDNTWN GP PA Total				
RTS-M, FT MCCOY, WI	052-62B10	(blank)	MOS	62b1
	091-45K40	2	ANCOC	45K4
	091-52D10	2	MOS	52d1
	091-52D30	2	BNCOC	52d3
	091-62B40	2	ANCOC	62B4
	091-63B/S/W10H8	(blank)	ASI	63BSWasi
	091-63B10	2	MOS	63b1
	091-63B30	2	BNCOC	63b3
	091-63B40	2	ANCOC	63B4
	091-63D/E/H/N/T/Y 10H8	(blank)	ASI	63DEHNT Yasi
	091-63D10	1	MOS	63d1
	091-63H10	2	MOS	63h1
	091-63H30	2	BNCOC	63h3
	091-63S10	2	MOS	63s1
	091-63W10	1	MOS	63w1
		2	MOS	63w1
	6-63-C42	(blank)	ANCOC	63-all
	610-63B10	(blank)	MOS	63b1
	610-63B30	(blank)	BNCOC	63b3

Schools Offering Specific Courses for the Various Options

Base Case		Nearest School		Reassign Courses Multifunctional		Reassign Courses Specialized		Consolidate Schools Specialized	
AC	RC	AC	RC	AC	RC	AC	RC	AC	RC
	9		9		17		27		41
	26				12				26
	5		9		15		23		30
	38		38				49		
	5		11		8				
	21		24		31		41		39
		11		11					
		5		5		6			
		9		9				70	
				5					
		7		7					
				40		40		40	
				5				9	
		5		5				7	
		5							
		24				8			
				5		8		7	
				5					
		3							
		5		5		24		16	
	242	**74**	**309**	**102**	**313**	**86**	**429**	**149**	**464**
	9		12		10				
	15		15		5				
	4		2		8		45		
	9		9		14				
	5		5		5				
	16		11		11				
	35		33		33		68		
	49		16		16				
	38		14		9				
	5		11		12				
	2		2		30				
	3		16		16				
	10		10		10				
	11		3		6				
	7		20		29		44		
	17		10		10				
				5					
							33		

School	Course ID	Phase	Course Level	MOS
	610-63W10	(blank)	MOS	63w1
	610-ASIH8 (63B/S)	(blank)	ASI	63B/S
	611-63D30	(blank)	BNCOC	63d3
	611-63H10	(blank)	MOS	63h1
	612-62B10	(blank)	MOS	62b1
	612-62B30	(blank)	BNCOC	62b3
	662-52D10	(blank)	MOS	52d1
	662-52D30	(blank)	BNCOC	52d3
	M1070 NET (63H1)	(blank)	New Eqpt	63H
	ULLS	(blank)	Other	?
RTS-M, FT MCCOY, WI Total				
RTS-M-01 SALINA KS	091-44B10	1	MOS	44b1
		2	MOS	44b1
		3	MOS	44b1
	091-44E10	1	MOS	44E10
		2	MOS	44E10
	091-44E30	2	BNCOC	44E30
		3	BNCOC	44E30
	091-45B10	1	MOS	45b1
		2	MOS	45b1
	091-52C10	1	MOS	52c1
		2	MOS	52c1
	091-52C30	2	BNCOC	52c3
	091-63B10	1	MOS	63b1
		2	MOS	63b1
	091-63B30	2	BNCOC	63b3
	091-63B40	2	ANCOC	63B4
	091-63D10	1	MOS	63d1
		2	MOS	63d1
	091-63D30 TRK II	2	BNCOC	63d3
	091-63D40	2	ANCOC	63D4
	091-63E10	1	MOS	63E10
		2	MOS	63E10
	091-63E40	2	ANCOC	63E40
	091-63G10	1	MOS	63g1
		2	MOS	63g1
	091-63H10	1	MOS	63h1
		2	MOS	63h1
	091-63H10 (GS) BFV	1	MOS	63h1
		2	MOS	63h1
	091-63H40	2	ANCOC	63H4

Schools Offering Specific Courses for the Various Options

Base Case		Nearest School		Reassign Courses Multifunctional		Reassign Courses Specialized		Consolidate Schools Specialized	
AC	RC	AC	RC	AC	RC	AC	RC	AC	RC
				5		46			
		249							
		6		5					
				48					
		12							
		4							
		20				17			
				16					
4		4		4		4			
13		13		11					
252	291	206	79	239	96	161			
27				24					
10		20		7		20		20	
6				9					
3		3							
1		1		1		1		1	
8		8		8					
8									
12		12		5		9		9	
5		12							
25				15				25	
45				30				42	
10				5					
19		9		9				22	
27		27		33		91		33	
5		14		21					
10		19		17					
34						38		38	
21						33		33	
12		12		7					
8		8							
4		9		21					
3		9		23					
10		16		5					
10		10		10		10		10	
11		11		6					
3		6		5		15		8	
33		18		15		44			
3		3		3		3		3	
5		5							
13		13		11		5		49	

106 Consolidating Active and Reserve Component Training Infrastructure

School	Course ID	Phase	Course Level	MOS
	091-63S10	1	MOS	63s1
		2	MOS	63s1
	091-63T10	1	MOS	63t1
		2	MOS	63t1
	091-63T40	2	ANCOC	63T4
	091-63W10	1	MOS	63w1
		2	MOS	63w1
	091-63W10 (GS) HEMTT	1	MOS	63w1
	091-63W10 (GS) HMMWV	1	MOS	63w1
		2	MOS	63w1
	091-63W10 (GS) M939	1	MOS	63w1
	091-63Y10	1	MOS	63y1
		2	MOS	63y1
	6-63-C42	(blank)	ANCOC	63-all
	610-63B30	(blank)	BNCOC	63b3
	610-63G10	(blank)	MOS	63g1
	610-63S10	(blank)	MOS	63s1
	610-63W10	(blank)	MOS	63w1
	611-63D10	(blank)	MOS	63d1
	611-63D30	(blank)	BNCOC	63d3
	611-63D30 (45D)	(blank)	BNCOC	63d3
	611-63E30	(blank)	BNCOC	63E30
	611-63H10	(blank)	MOS	63h1
	611-63T10	(blank)	MOS	63t1
	611-63T30	(blank)	BNCOC	63t3
	611-63Y10	(blank)	MOS	63y1
	641-45B10	(blank)	MOS	45b1
	662-52C10	(blank)	MOS	52c1
	662-52C30	(blank)	BNCOC	52c3
	662-52C30 (63J)	(blank)	BNCOC	52c3
	702-44E10	(blank)	MOS	44E10
	702-44E30	(blank)	BNCOC	44E30
	702-44E30 (44B)	(blank)	BNCOC	44E30 (44B)
	704-44B10	(blank)	MOS	44b1
RTS-M-01 SALINA KS Total				
RTS-M-02 CAMP DODGE IA	052-62B10	(blank)	MOS	62b1
	052-62B30	2	BNCOC	62b3
	091-45B10	2	MOS	45b1
	091-45K10	1	MOS	45k1

Schools Offering Specific Courses for the Various Options

Base Case		Nearest School		Reassign Courses Multifunctional		Reassign Courses Specialized		Consolidate Schools Specialized	
AC	RC	AC	RC	AC	RC	AC	RC	AC	RC
	19		9		9		30		
	19		22		17		37		
	12		35		9				24
	26		23		23		28		19
	11				6				
	24		31		14		38		
	27		19		12		13		9
	12				7				
	3		3						
	3		3						
	8		8						
	12		9		5				
	6		5		5				
		65		65		5		5	
		14		15		5		17	
		2							
				17					
		24							
				8					
				16		19		19	
						15			
								19	
				16				6	
								8	
				5					
						24		24	
				5		15		16	
						12			
		1		1					
				8		17		17	
				14		21		21	
		28		12		11		7	
	573	**134**	**412**	**182**	**397**	**144**	**415**	**159**	**345**
	36		6		5				
	14		7		6				
	4				7				
	47		45		26				

School	Course ID	Phase	Course Level	MOS
		2	MOS	45k1
		3	MOS	45k1
	091-45K30	2	BNCOC	45k3
		3	BNCOC	45k3
	091-52C10	1	MOS	52c1
	091-52D10	1	MOS	52d1
		2	MOS	52d1
	091-52D30	2	BNCOC	52d3
	091-63B/S/W10H8	(blank)	ASI	63BSWasi
	091-63B10	1	MOS	63b1
		2	MOS	63b1
	091-63B30	2	BNCOC	63b3
	091-63B40	2	ANCOC	63B4
	091-63D/E/H/N/T/Y10H8	(blank)	ASI	63DEHNTYasi
	091-63E10	2	MOS	63E10
	091-63H10	1	MOS	63h1
		2	MOS	63h1
	091-63H10 (GS) M1	(blank)	MOS	63h1
	091-63H10L8	(blank)	ASI	63H
	091-63H30-IIA	2	BNCOC	63h3
	091-63H30-IIB	2	BNCOC	63h3
	091-63H40	2	ANCOC	63H4
	091-63J10	1	MOS	63j1
	091-63S10	1	MOS	63s1
	091-63W10	1	MOS	63w1
		2	MOS	63w1
	091-63W10 (GS) M939	(blank)	MOS	63w1
	093-27E10	1	MOS	27e
		2	MOS	27e
		3	MOS	27e
		4	MOS	27e
	093-27E30	2	BNCOC	27e
	6-63-C42	(blank)	ANCOC	63-all
	610-63B10	(blank)	MOS	63b1
	610-63B30	(blank)	BNCOC	63b3
	610-63S10	(blank)	MOS	63s1
	610-63W10	(blank)	MOS	63w1
	610-ASIH8 (63B/S)	(blank)	ASI	63B/S
	611-63E10	(blank)	MOS	63E10
	611-63E30	(blank)	BNCOC	63E30

Schools Offering Specific Courses for the Various Options

Base Case		Nearest School		Reassign Courses Multifunctional		Reassign Courses Specialized		Consolidate Schools Specialized	
AC	RC	AC	RC	AC	RC	AC	RC	AC	RC
	30				27				
	14				5				
	10		3		5		16		16
	7		10		5		17		17
	21				6		66		
	44		14		13		34		39
	25		9		7				
	9		5						
	30		22		20		60		
	17		14		14		33		
	15		20		23		64		23
	27		10		11		43		52
	12		12		11				
	5		5		9				
	9				9				
	49		12		10				
	52		5		7				49
	8								
	6		6						
	27				14		38		
	28		28		17				
	23		9		11				
	50				28				31
	2		5		5				
	85		30		6				
	78		49		28				
	8		8		8		8		8
	20		20		20		20		20
	18		18		18		18		18
	9		9		9		9		9
	9		9		9		9		9
	6		6		6		6		6
		5						45	
		39		38				14	
		7		23					
		7							
				17					
				37		39			
				6		25		25	

School	Course ID	Phase	Course Level	MOS
	611-63H10	(blank)	MOS	63h1
	611-63H30	(blank)	BNCOC	63h3
	641-45B10	(blank)	MOS	45b1
	643-45K10	(blank)	MOS	45k1
	643-45K30	(blank)	BNCOC	45k3
	662-52C10	(blank)	MOS	52c1
	662-52C30	(blank)	BNCOC	52c3
	662-52C30 (63J)	(blank)	BNCOC	52c3
	662-52D10	(blank)	MOS	52d1
	662-52D30	(blank)	BNCOC	52d3
	690-63J10	(blank)	MOS	63j1
	TAMMS	(blank)	Other	?
	ULLS-G (S)	(blank)	Sustainment	?
RTS-M-02 CAMP DODGE IA Total				
RTS-M-03 FT DIX NJ	091-52D10	1	MOS	52d1
		2	MOS	52d1
	091-52D30	2	BNCOC	52d3
	091-63B/S/W10H8	(blank)	ASI	63BSWasi
	091-63B10	1	MOS	63b1
		2	MOS	63b1
	091-63B30	2	BNCOC	63b3
	091-63B40	2	ANCOC	63B4
	091-63D/E/H/N/T/Y10H8	(blank)	ASI	63DEHNTYasi
	091-63H10	1	MOS	63h1
		2	MOS	63h1
	091-63H10 (T)	(blank)	Transition	63H
	091-63S10	1	MOS	63s1
	091-63T10	1	MOS	63t1
		2	MOS	63t1
	091-63W10	1	MOS	63w1
		2	MOS	63w1
	091-63Y10	1	MOS	63y1
		2	MOS	63y1
	091-ASIL8 (45K)	1	ASI	45Kasi
	101-92A10	1	MOS	92a1
		2	MOS	92a1
	551-92A10	(blank)	MOS	92a1
	551-92A30	(blank)	BNCOC	92a3
	610-63B10	(blank)	MOS	63b1
	610-63B30	(blank)	BNCOC	63b3

Schools Offering Specific Courses for the Various Options

Base Case		Nearest School		Reassign Courses Multifunctional		Reassign Courses Specialized		Consolidate Schools Specialized	
AC	RC	AC	RC	AC	RC	AC	RC	AC	RC
						42			
						35		57	
		1		5					
				5					
				19		12		12	
		7		5					
				5				9	
				7					
		38		8					
		5							
				14		12		8	
8		12		5					
101		72		41					
963	109	480	189	451	165	441	170	297	
11		7		5					
9		7		5					14
6		5		5					
12		5		5		47			47
11		20		20					37
33		32		23		56			56
32		24		24		54			54
17		14		14					
18		18			7	56			56
7		7			7	22			22
4		12		5					
10					11	11			11
2		3			12				
14		14			11	49			58
33		33			71	22			72
12		7			7	20			
16		12		5					
21		19			10	49			49
15		8			8	50			50
6		6			6				
46		113			47	52			54
79		70			70	110			79
			5		33				
			63		16				
								5	
							8		

School	Course ID	Phase	Course Level	MOS
	610-63W10	(blank)	MOS	63w1
	611-63H10	(blank)	MOS	63h1
	611-63T10	(blank)	MOS	63t1
	611-63T30	(blank)	BNCOC	63t3
	611-63Y10	(blank)	MOS	63y1
	662-52D10	(blank)	MOS	52d1
	662-52D30	(blank)	BNCOC	52d3
	TAMMS	(blank)	Other	?
	ULLS-G (S)	(blank)	Sustainment	?
RTS-M-03 FT DIX NJ Total				
RTS-M-04 FT BRAGG NC	052-62B10	(blank)	MOS	62b1
	052-62B30	2	BNCOC	62b3
	091-45K10	3	MOS	45k1
	091-45K30	2	BNCOC	45k3
		3	BNCOC	45k3
	091-52D10	1	MOS	52d1
		2	MOS	52d1
	091-52D30	2	BNCOC	52d3
	091-63B/S/W10H8	(blank)	ASI	63BSWasi
	091-63B10	1	MOS	63b1
		2	MOS	63b1
	091-63B30	2	BNCOC	63b3
	091-63H10	1	MOS	63h1
		2	MOS	63h1
	091-63H30	2	BNCOC	63h3
	091-63S10	1	MOS	63s1
		2	MOS	63s1
	091-63T10	1	MOS	63t1
		2	MOS	63t1
	091-63T30	2	BNCOC	63t3
	091-63W10	1	MOS	63w1
		2	MOS	63w1
	101-92A10	1	MOS	92a1
		2	MOS	92a1
	551-92A10	(blank)	MOS	92a1
	551-92A30	(blank)	BNCOC	92a3
	6-63-C42	(blank)	ANCOC	63-all
	610-63B10	(blank)	MOS	63b1
	610-63B30	(blank)	BNCOC	63b3
	610-63S10	(blank)	MOS	63s1
	610-63W10	(blank)	MOS	63w1

Schools Offering Specific Courses for the Various Options

Base Case		Nearest School		Reassign Courses Multifunctional		Reassign Courses Specialized		Consolidate Schools Specialized	
AC	RC	AC	RC	AC	RC	AC	RC	AC	RC
		6							
						8			
						25			
				5				25	
				5					
22									
57		57		60					
493	74	493	59	438	41	598		30	659
47			36		30		50		44
6			7	5					
8				5			26		26
3			10	5					
3			5	5					
9				17		56			65
12			7	7					39
6			6	6					58
15			15	12					
20			22	22					
23			18	18					
25			16	10		10			10
12			15	28					
4			11	10					10
4			4	5					
13			22	14					45
16			14	8					
27				37					
8			9	9		20			
7			10	5					
14			14	19		81			
9			9	14		14			22
26			64	34		39			107
35			43	43					43
			48	48					
			47			15			
			83	74		64			55
			22			46			
			48	11					26
			3						
			7						19

School	Course ID	Phase	Course Level	MOS
	610-ASIH8 (63B/S)	(blank)	ASI	63B/S
	611-63H10	(blank)	MOS	63h1
	611-63T10	(blank)	MOS	63t1
	611-63T30	(blank)	BNCOC	63t3
	612-62B10	(blank)	MOS	62b1
	612-62B30	(blank)	BNCOC	62b3
	643-45K10	(blank)	MOS	45k1
	643-45K30	(blank)	BNCOC	45k3
	662-52D10	(blank)	MOS	52d1
	662-52D30	(blank)	BNCOC	52d3
	ULLS-G (S)	(blank)	Sustainment	?
RTS-M-04 FT BRAGG NC Total				
RTS-M-05 CAMP SHELBY MS	052-62B10	(blank)	MOS	62b1
	052-62B30	2	BNCOC	62b3
	091-52D10	2	MOS	52d1
	091-52D30	2	BNCOC	52d3
	091-63B10	2	MOS	63b1
	091-63B30	2	BNCOC	63b3
	091-63B40	2	ANCOC	63B4
	091-63D10	2	MOS	63d1
	091-63H10	1	MOS	63h1
		2	MOS	63h1
	091-63H30-IIA	2	BNCOC	63h3
	091-63H40	2	ANCOC	63H4
	091-63S10	1	MOS	63s1
		2	MOS	63s1
	091-63T10	2	MOS	63t1
	101-92A10	2	MOS	92a1
	551-92A10	(blank)	MOS	92a1
	551-92A30	(blank)	BNCOC	92a3
	6-63-C42	(blank)	ANCOC	63-all
	610-63B10	(blank)	MOS	63b1
	610-63B30	(blank)	BNCOC	63b3
	610-63S10	(blank)	MOS	63s1
	611-63D10	(blank)	MOS	63d1
	611-63H10	(blank)	MOS	63h1
	611-63H30	(blank)	BNCOC	63h3
	611-63T10	(blank)	MOS	63t1
	611-63T30	(blank)	BNCOC	63t3
	612-62B10	(blank)	MOS	62b1
	662-52D10	(blank)	MOS	52d1

Schools Offering Specific Courses for the Various Options   115

Base Case		Nearest School		Reassign Courses Multifunctional		Reassign Courses Specialized		Consolidate Schools Specialized	
AC	RC	AC	RC	AC	RC	AC	RC	AC	RC
				23					
		18				5			
				5		23		5	
								15	
				8		25		27	
				7					
						8		8	
		10				6		6	
						18			
		6							
36		38		38		38		38	
388	245	395	223	406	210	334	161	507	
7		21		21					
10		7		5					
13		15		9		50			42
11		13		13					
54		32		32					32
40		46		37		44			29
51		30		22		50			17
9		25		14					
10		34		35		87			87
28		36		14		14			23
6		26		9					38
14		7		7					
9		3		5					
16		10		10					
7		10		10					
100		40		40					
		57		57		57		57	
				54					
		12		12		12		12	
		5		5				27	
		28		13		65		14	
		7							
		8							
								44	
				33		21			
		13		6		6		27	
						15			
				5					
				7				17	

School	Course ID	Phase	Course Level	MOS
	662-52D30	(blank)	BNCOC	52d3
	TAMMS	(blank)	Other	?
	ULLS-G (S)	(blank)	Sustainment	?
RTS-M-05 CAMP SHELBY MS Total				
RTS-M-06 CAMP ROBERT CA	052-62B10	(blank)	MOS	62b1
	091-45E10	1	MOS	45E10
		2	MOS	45E10
	091-45K10 (T)	(blank)	Transition	45K
	091-52D10	1	MOS	52d1
		2	MOS	52d1
	091-63B/S/W10H8	(blank)	ASI	63BSWasi
	091-63B10	1	MOS	63b1
		2	MOS	63b1
	091-63B30	2	BNCOC	63b3
	091-63B40	2	ANCOC	63B4
	091-63E10	1	MOS	63E10
		2	MOS	63E10
	091-63H10	1	MOS	63h1
		2	MOS	63h1
	091-63H10 (T)	(blank)	Transition	63H
	091-63H30	2	BNCOC	63h3
	091-63H40	2	ANCOC	63H4
	091-63S10	1	MOS	63s1
		2	MOS	63s1
	091-63T10	1	MOS	63t1
		2	MOS	63t1
	091-63W10	1	MOS	63w1
		2	MOS	63w1
	101-92A10	1	MOS	92a1
		2	MOS	92a1
	551-92A10	(blank)	MOS	92a1
	551-92A30	(blank)	BNCOC	92a3
	6-63-C42	(blank)	ANCOC	63-all
	610-63B30	(blank)	BNCOC	63b3
	610-63S10	(blank)	MOS	63s1
	610-63W10	(blank)	MOS	63w1
	610-ASIH8 (63B/S)	(blank)	ASI	63B/S
	611-63E10	(blank)	MOS	63E10
	611-63E30	(blank)	BNCOC	63E30
	611-63H10	(blank)	MOS	63h1
	611-63T10	(blank)	MOS	63t1

Schools Offering Specific Courses for the Various Options

Base Case		Nearest School		Reassign Courses Multifunctional		Reassign Courses Specialized		Consolidate Schools Specialized	
AC	RC	AC	RC	AC	RC	AC	RC	AC	RC
		8				12			
7				5		33		33	
6		6		12					
398	138	361	192	300	188	278		198	301
7		6		6					
4		4		4		4		4	
3				9		9		9	
14		14		14					
24		22		22					
18		17		17		21		21	
13		13		15		24		15	
57		41		41		41		41	
64		45		45					
27		18		18		18		18	
14		14		14				14	
21		27				20		20	
23						23		23	
25		30		36		36		36	
20		27		27					
49		48		48		48		48	
4		4		5					
9		5		5					
15		15		15		25		25	
10		11		11		11		11	
16		20		7		8			
10		6		6		18		6	
15		28		35		32		35	
15		28		32		32		32	
42		53		46		44		44	
39		50		50		49		49	
		14		14		36		36	
		41		34		34		34	
		16		16		40		45	
		12		9		5		5	
		3		5		6		6	
		8		8		15		15	
		32		32		32		32	
		5		5					
		5		5					
				5		5			
		5		5				5	

School	Course ID	Phase	Course Level	MOS
	611-63T30	(blank)	BNCOC	63t3
	612-62B10	(blank)	MOS	62b1
	612-62B30	(blank)	BNCOC	62b3
	643-45E10	(blank)	MOS	45E10
	643-45K10	(blank)	MOS	45k1
	662-52D10	(blank)	MOS	52d1
	662-52D30	(blank)	BNCOC	52d3
	ULLS-G (S)	(blank)	Sustainment	?
RTS-M-06 CAMP ROBERT CA Total				
RTS-M-07 CAMP RIPLEY MN	052-62B10	(blank)	MOS	62b1
	091-45D10	1	MOS	45d1
		2	MOS	45d1
	091-45E10	2	MOS	45E10
	091-63B/S/W10H8	(blank)	ASI	63BSWasi
	091-63B10	2	MOS	63b1
	091-63B30	2	BNCOC	63b3
	091-63B40	2	ANCOC	63B4
	091-63D/E/H/N/T/Y10H8	(blank)	ASI	63DEHNTYasi
	091-63D10	2	MOS	63d1
	091-63E10	2	MOS	63E10
	091-63H10	2	MOS	63h1
	091-63H30	2	BNCOC	63h3
	091-63H40	2	ANCOC	63H4
	091-63S10	2	MOS	63s1
	091-63T10	2	MOS	63t1
	091-63T30	2	BNCOC	63t3
	091-63W10	2	MOS	63w1
	101-92A10	2	MOS	92a1
	101-92A30	2	BNCOC	92a3
	101-92A40	2	ANCOC	92a4
	551-92A10	(blank)	MOS	92a1
	551-92A30	(blank)	BNCOC	92a3
	551-92A40	(blank)	ANCOC	92a4
	6-63-C42	(blank)	ANCOC	63-all
	610-63B30	(blank)	BNCOC	63b3
	610-63W10	(blank)	MOS	63w1
	611-63D10	(blank)	MOS	63d1
	611-63D30	(blank)	BNCOC	63d3
	611-63E10	(blank)	MOS	63E10
	611-63E30	(blank)	BNCOC	63E30

Schools Offering Specific Courses for the Various Options

Base Case		Nearest School		Reassign Courses Multifunctional		Reassign Courses Specialized		Consolidate Schools Specialized	
AC	RC	AC	RC	AC	RC	AC	RC	AC	RC
		6		5		6		6	
		12		5					
		27		5					
				8					
		13							
		26		6		8		6	
		8		8		10		12	
20		20		20					
578	233	566	175	548	197	463	202	451	
6		14		14				48	
11				6					
20				14					
6		9							
10		9		9				52	
43		16		31				45	
23		29		27					
15		16		36		101		71	
11		10		5					
32						49		49	
16				16					
22		21		28		63			
11		11		10					
6		15		17					
9		14		14				55	
18		20		20		20		23	
3		13		5		23		23	
41		51		45		82		82	
39		39		51		51		51	
15		27		12					
7				9					
				28					
		40		46					
		20		21		48		47	
						103			
						5		19	
				14				27	
						8		8	
		6							
				5		11		7	

120 Consolidating Active and Reserve Component Training Infrastructure

School	Course ID	Phase	Course Level	MOS
	611-63T10	(blank)	MOS	63t1
	611-63T30	(blank)	BNCOC	63t3
	612-62B10	(blank)	MOS	62b1
	642-45D10	(blank)	MOS	45d1
	643-45E10	(blank)	MOS	45E10
	ULLS-G (S)	(blank)	Sustainment	?
RTS-M-07 CAMP RIPLEY MN Total				
RTS-M-08 FT CUSTER MI	052-62B10	(blank)	MOS	62b1
	091-52D10	1	MOS	52d1
		2	MOS	52d1
	091-63B10	1	MOS	63b1
		2	MOS	63b1
	091-63B30	2	BNCOC	63b3
	091-63D/E/H/N/T/Y10H8	(blank)	ASI	63DEHNT Yasi
	091-63E10	1	MOS	63E10
		2	MOS	63E10
	091-63H10	1	MOS	63h1
		2	MOS	63h1
	091-63S10	1	MOS	63s1
		2	MOS	63s1
	091-63T10	1	MOS	63t1
		2	MOS	63t1
	091-63Y10	1	MOS	63y1
		2	MOS	63y1
	6-63-C42	(blank)	ANCOC	63-all
	610-63B10	(blank)	MOS	63b1
	610-63B30	(blank)	BNCOC	63b3
	610-63S10	(blank)	MOS	63s1
	611-63E10	(blank)	MOS	63E10
	611-63E30	(blank)	BNCOC	63E30
	611-63H10	(blank)	MOS	63h1
	611-63T10	(blank)	MOS	63t1
	611-63Y10	(blank)	MOS	63y1
	612-62B10	(blank)	MOS	62b1
	612-62B30	(blank)	BNCOC	62b3
	662-52D10	(blank)	MOS	52d1
	662-52D30	(blank)	BNCOC	52d3
RTS-M-08 FT CUSTER MI Total				
RTS-M-09 GOWEN FIELD ID	052-62B10	(blank)	MOS	62b1
	052-62B30	2	BNCOC	62b3

Schools Offering Specific Courses for the Various Options

Base Case		Nearest School		Reassign Courses Multifunctional		Reassign Courses Specialized		Consolidate Schools Specialized	
AC	RC	AC	RC	AC	RC	AC	RC	AC	RC
		5		5					
		14						26	
						14		14	
								8	
	17		19		23				
	381	85	333	119	392	189	389	156	499
	11		11		23				
	4		17		8				
	10		17		16				
	32		32		42		59		53
	39		52		56		56		75
	10		14		39		72		66
	7		7		7				
	11				10				22
	7				5				
	41		33		36		39		46
	30		23		23				51
	16		25		27				
	15		20		28		44		
	8		14		10		32		
	7		6		6		11		
	12		12		16				
	8		7		8				
		5		45				56	
				12					
		5		40		40		50	
				5					
								17	
				5		31			
		14		29		17			
				5				25	
				5					
				5					
				5		19		19	
	268	24	290	161	360	107	313	167	313
	22		21		21		27		27
	3		10		10		13		13

School	Course ID	Phase	Course Level	MOS
	091-45B10	(blank)	MOS	45b1
	091-45E10	(blank)	MOS	45E10
	091-45K10	(blank)	MOS	45k1
	091-62B40	2	ANCOC	62B4
	091-63B10	1	MOS	63b1
		2	MOS	63b1
	091-63B30	2	BNCOC	63b3
	091-63B40	2	ANCOC	63B4
	091-63E10	(blank)	MOS	63E10
	091-63E30	2	BNCOC	63E30
	091-63E40	2	ANCOC	63E40
	091-63H10	(blank)	MOS	63h1
	091-63H30	2	BNCOC	63h3
	091-63H40	2	ANCOC	63H4
	091-63S10	1	MOS	63s1
		2	MOS	63s1
	091-63T10	1	MOS	63t1
		2	MOS	63t1
	091-63T30	2	BNCOC	63t3
	091-63W10	(blank)	MOS	63w1
	091-63Y10	2	MOS	63y1
	6-63-C42	(blank)	ANCOC	63-all
	610-63B10	(blank)	MOS	63b1
	610-63B30	(blank)	BNCOC	63b3
	610-63S10	(blank)	MOS	63s1
	610-63W10	(blank)	MOS	63w1
	611-63E10	(blank)	MOS	63E10
	611-63E30	(blank)	BNCOC	63E30
	611-63H10	(blank)	MOS	63h1
	611-63T10	(blank)	MOS	63t1
	611-63T30	(blank)	BNCOC	63t3
	611-63Y10	(blank)	MOS	63y1
	612-62B10	(blank)	MOS	62b1
	612-62B30	(blank)	BNCOC	62b3
	641-45B10	(blank)	MOS	45b1
	643-45E10	(blank)	MOS	45E10
	643-45K10	(blank)	MOS	45k1
	643-45K30	(blank)	BNCOC	45k3
	ULLS-G (S)	(blank)	Sustainment	?
RTS-M-09 GOWEN FIELD ID Total				
RTS-M-10 BLANDING FL	091-45B10	(blank)	MOS	45b1

Schools Offering Specific Courses for the Various Options   123

Base Case		Nearest School		Reassign Courses Multifunctional		Reassign Courses Specialized		Consolidate Schools Specialized	
AC	RC	AC	RC	AC	RC	AC	RC	AC	RC
	4		4		5				
	25				12		25		25
	7		7		7		7		7
	7		7		6		16		16
	32		36		36		36		36
	32		36		36				81
	14		18		18		27		18
	8		15		15				21
	39				16		39		39
	6				7				
	13		7		5		23		23
	23				12				
	4		4		5				
	1		5		5		64		9
	25		12		12				
	22		11		11		11		12
	17		24		19		18		25
	17		16		16				16
	8				6				
	21		21		21		21		
	3		5		5				
		36		36					
		5		16		45		45	
		24		37		36		34	
		14		5					
		11		17					
		5		5					
		5		5					
				5				20	
		22		7		5			
				5					
		12							
		28		5		8		6	
				6		12		14	
		12		5					
		8							
						5		5	
				13		9		9	
	55		55		72		330		330
	408	182	314	167	378	120	657	133	698
	16				10				

School	Course ID	Phase	Course Level	MOS
	091-45K10	2	MOS	45k1
	091-45K40	2	ANCOC	45K4
	091-52D10	2	MOS	52d1
		(blank)	MOS	52d1
	091-63B/S/W10H8	(blank)	ASI	63BSWasi
	091-63B10	1	MOS	63b1
		2	MOS	63b1
	091-63B30	2	BNCOC	63b3
	091-63B40	2	ANCOC	63B4
	091-63W10	2	MOS	63w1
	101-92A10	2	MOS	92a1
	551-92A10	(blank)	MOS	92a1
	551-92A30	(blank)	BNCOC	92a3
	6-63-C42	(blank)	ANCOC	63-all
	610-63B10	(blank)	MOS	63b1
	610-63B30	(blank)	BNCOC	63b3
	610-63W10	(blank)	MOS	63w1
	610-ASIH8 (63B/S)	(blank)	ASI	63B/S
	641-45B10	(blank)	MOS	45b1
	643-45K30	(blank)	BNCOC	45k3
	662-52D10	(blank)	MOS	52d1
	662-52D30	(blank)	BNCOC	52d3
	ULLS-G (S)	(blank)	Sustainment	?
RTS-M-10 BLANDING FL Total				
RTS-M-12 FT STEWART GA	052-62B10 (T)	(blank)	Transition	62B
	052-62B30	2	BNCOC	62b3
	091-44B10	1	MOS	44b1
		2	MOS	44b1
		3	MOS	44b1
	091-45K10	1	MOS	45k1
		2	MOS	45k1
		3	MOS	45k1
	091-45T10	1	MOS	45t1
		2	MOS	45t1
	091-52C10	1	MOS	52c1
	091-52D10	1	MOS	52d1
		2	MOS	52d1
	091-63B10	1	MOS	63b1
		2	MOS	63b1
	091-63D10	1	MOS	63d1
		2	MOS	63d1

Schools Offering Specific Courses for the Various Options   125

Base Case		Nearest School		Reassign Courses Multifunctional		Reassign Courses Specialized		Consolidate Schools Specialized	
AC	RC	AC	RC	AC	RC	AC	RC	AC	RC
	6		6						
	1		1		5		16		16
	7		1						
	11		11		6				
	11		16		12				
	11		10		10				37
	11		9		9		25		29
	4		4		13		13		34
	6		6		16		16		29
	21		40		18		41		41
	49		49		49		103		93
					24		24		24
		56			30		37		56
		5			68		70		59
		5			44		20		
					12		16		16
					49				
					5				
					6		14		14
					18		25		11
		5			7				11
	81		81		84		98		98
	235	71	234	263	232	206	312	191	377
	2		2						2
	13		9		7		24		
	18				14				
	10				8				
	3		9						
	6				18		49		49
	6		17		11		36		36
	4				11				
	6		6		6		6		6
	2		2		2		2		2
	11		11						41
	8		8		28				
	1		1		24				
	26		11		11				
	29		47		32				
	13		42		14				
	15				9		25		25

School	Course ID	Phase	Course Level	MOS
		(blank)	MOS	63d1
	091-63E10	1	MOS	63E10
	091-63H10	1	MOS	63h1
		2	MOS	63h1
	091-63H30	2	BNCOC	63h3
	091-63H40	2	ANCOC	63H4
	091-63J10	1	MOS	63j1
	091-63S10	1	MOS	63s1
		2	MOS	63s1
	091-63T10	1	MOS	63t1
		2	MOS	63t1
	091-63W10	1	MOS	63w1
		2	MOS	63w1
	091-63Y10	1	MOS	63y1
		2	MOS	63y1
	101-92A10	1	MOS	92a1
		2	MOS	92a1
	101-92A30	2	BNCOC	92a3
	101-92A40	2	ANCOC	92a4
	551-92A10	(blank)	MOS	92a1
	551-92A30	(blank)	BNCOC	92a3
	551-92A40	(blank)	ANCOC	92a4
	6-63-C42	(blank)	ANCOC	63-all
	610-63B10	(blank)	MOS	63b1
	610-63B30	(blank)	BNCOC	63b3
	610-63S10	(blank)	MOS	63s1
	610-63W10	(blank)	MOS	63w1
	611-63D30	(blank)	BNCOC	63d3
	611-63D30 (45D)	(blank)	BNCOC	63d3
	611-63E10	(blank)	MOS	63E10
	611-63E30	(blank)	BNCOC	63E30
	611-63H10	(blank)	MOS	63h1
	611-63T10	(blank)	MOS	63t1
	611-63T30	(blank)	BNCOC	63t3
	612-62B30	(blank)	BNCOC	62b3
	643-45K30	(blank)	BNCOC	45k3
	662-52C10	(blank)	MOS	52c1
	662-52C30	(blank)	BNCOC	52c3
	662-52C30 (63J)	(blank)	BNCOC	52c3
	662-52D30	(blank)	BNCOC	52d3
	690-63J10	(blank)	MOS	63j1

Schools Offering Specific Courses for the Various Options

Base Case		Nearest School		Reassign Courses Multifunctional		Reassign Courses Specialized		Consolidate Schools Specialized	
AC	RC	AC	RC	AC	RC	AC	RC	AC	RC
	20				12				
	6	6		6			22		
	38	50							
	33	50		62			68		54
	14	14		12					
	12	14		14					
	13	13		22			45		
	16	20		14					
	20	29		14			34		46
	13			9					
	23	21		21			17		44
	28			69					75
	42			88			59		59
	4	9		5					
	11	11		10					
	123			82			71		
	77	104		104			85		94
	6	6		16					
	2	2		5			24		24
		43							
				5					
				7					
		94				5			
		19							
		70				20		17	
		7							
		31							
		5		5		15		15	
		7		7		7		7	
		10				9			
		9				5			
				13					
				11				17	
		5							
				5		25		31	
						6		6	
				6					
		20		5		10			
		24		5					
		5							
				8		10		22	

School	Course ID	Phase	Course Level	MOS
	704-44B10	(blank)	MOS	44b1
	ULLS-G (S)	(blank)	Sustainment	?
RTS-M-12 FT STEWART GA Total				
USATC, FT. JACKSON/108TH	091-63B10	1	MOS	63b1
		2	MOS	63b1
	091-63S10	1	MOS	63s1
		2	MOS	63s1
	610-63B10	(blank)	MOS	63b1
	610-63S10	(blank)	MOS	63s1
	610-ASIH8 (63B/S)	(blank)	ASI	63B/S
USATC, FT. JACKSON/108TH Total				
USATC, FT. WOOD/98TH DIV	052-62B10	(blank)	MOS	62b1
	052-62B10 (T)	(blank)	Transition	62B
	052-62B40	2	ANCOC	62B4
	612-62B10	(blank)	MOS	62b1
USATC, FT. WOOD/98TH DIV Total				
Waiwa RTSM	091-63B/S/W10H8	(blank)	ASI	63BSWasi
	091-63B10	2	MOS	63b1
	101-92A10	1	MOS	92a1
		2	MOS	92a1
	101-92A30	2	BNCOC	92a3
	551-92A10	(blank)	MOS	92a1
	551-92A30	(blank)	BNCOC	92a3
	551-92A40	(blank)	ANCOC	92a4
	6-63-C42	(blank)	ANCOC	63-all
	610-63B10	(blank)	MOS	63b1
	610-63B30	(blank)	BNCOC	63b3
	610-ASIH8 (63B/S)	(blank)	ASI	63B/S
	TAMMS	(blank)	Other	?
	ULLS-G (S)	(blank)	Sustainment	?
Waiwa RTSM Total				

Schools Offering Specific Courses for the Various Options   129

Base Case		Nearest School		Reassign Courses Multifunctional		Reassign Courses Specialized		Consolidate Schools Specialized	
AC	RC	AC	RC	AC	RC	AC	RC	AC	RC
				5		17		21	
	38		41		34				
712	349	555	82	794	129	567	136	557	
			10		10		57		
			9		20		70		50
			3		5		68		
			3		10				
159			18		47			59	
52			7		8		38		
286							98		98
497		25	25	55	45	136	195	157	50
			21		27		34		35
					2		2		
			16		16		16		16
66				33					
66			37	33	45		52		51
	9		9		9				9
	6		6		6		6		6
	13		11		11		11		11
	28		25		25		25		25
	5		5		5				
		22		22					
		12		12		12		12	
		115		7		7		7	
		13		13		13		13	
		9		9		9		9	
		12		12		17		16	
		5		5		5		5	
	1		1		5		5		5
	36		33		33				
98	188	90	80	94	63	47	62	56	

REFERENCES

Shanley, Michael G., John D. Winkler, and Paul S. Steinberg, *Resources, Costs, and Efficiency of Training in the Total Army School System*, Santa Monica, CA: RAND, MR-844-A, 1997.

Winkler, John D., Michael G. Shanley, James C. Crowley, Rodger Madison, Diane Green, J. Michael Polich, Paul S. Steinberg, and Laurie L. McDonald, *Assessing the Performance of the Army Reserve Components School System*, Santa Monica, CA: RAND, MR-590-A, 1996.

Winkler, John D., John Schank, Michael Mattock, Rodger A. Madison, Diane Green, James C. Crowley, Laurie L. McDonald, and Paul S. Steinberg, *Training Requirements and Training Delivery in the Total Army School System*, Santa Monica, CA: RAND, MR-928-A, forthcoming.

Winkler, John D., Michael G. Shanley, John Schank, James C. Crowley, Michael Mattock, Diane Green, Rodger A. Madison, Laurie L. McDonald, and Paul S. Steinberg, *The Total Army School System: Recommendations for Future Policy*, Santa Monica, CA: RAND, MR-955-A, forthcoming.